An Introduction to the Planning Domain Definition Language

Praise for *An Introduction to the Planning Domain Definition Language*

This is a thorough and comprehensive introduction to PDDL. It does an excellent job of demonstrating PDDL's capabilities and will be an essential resource for anyone who wants to use PDDL. It will also be an important resource for anyone who studies AI planning, regardless of whether they ever use PDDL. A book like this has been needed for a long time.

–Dana Nau, *University of Maryland*

Synthesis Lectures on Artificial Intelligence and Machine Learning

Editors
Ronald J. Brachman, *Jacobs Technion-Cornell Institute at Cornell Tech*
Francesca Rossi, *IBM Research AI*
Peter Stone, *University of Texas at Austin*

An Introduction to the Planning Domain Definition Language
Patrik Haslum, Nir Lipovetzky, Daniele Magazzeni, and Christian Muise
2019

Reasoning with Probabilistic and Deterministic Graphical Models: Exact Algorithms,
Second Edition
Rina Dechter
2019

Learning and Decision-Making from Rank Data
Liron Xia
2019

Lifelong Machine Learning, Second Edition
Zhiyuan Chen and Bing Liu
2018

Adversarial Machine Learning
Yevgeniy Vorobeychik and Murat Kantarcioglu
2018

Strategic Voting
Reshef Meir
2018

Predicting Human Decision-Making: From Prediction to Action
Ariel Rosenfeld and Sarit Kraus
2018

An Introduction to the Planning Domain Definition Language
Patrik Haslum, Nir Lipovetzky, Daniele Magazzeni, and Christian Muise

ISBN: 978-3-031-00456-8   paperback
ISBN: 978-3-031-01584-7   ebook
ISBN: 978-3-031-00029-4   hardcover

DOI 10.1007/978-3-031-01584-7

A Publication in the Springer series
SYNTHESIS LECTURES ON ARTIFICIAL INTELLIGENCE AND MACHINE LEARNING

Lecture #42
Series Editors: Ronald J. Brachman, *Jacobs Technion_Cornell Institute at Cornell Tech*
                Francesca Rossi, *IBM Research AI*
                Peter Stone, *University of Texas at Austin*
Series ISSN
Synthesis Lectures on Artificial Intelligence and Machine Learning
Print 1939-4608   Electronic 1939-4616

An Introduction to the Planning Domain Definition Language

Patrik Haslum
Australian National University

Nir Lipovetzky
University of Melbourne

Daniele Magazzeni
King's College London

Christian Muise
IBM Research

SYNTHESIS LECTURES ON ARTIFICIAL INTELLIGENCE AND MACHINE LEARNING #42

ABSTRACT

Planning is the branch of Artificial Intelligence (AI) that seeks to automate reasoning about plans, most importantly the reasoning that goes into formulating a plan to achieve a given goal in a given situation. AI planning is model-based: a planning system takes as input a description (or model) of the initial situation, the actions available to change it, and the goal condition to output a plan composed of those actions that will accomplish the goal when executed from the initial situation.

The Planning Domain Definition Language (PDDL) is a formal knowledge representation language designed to express planning models. Developed by the planning research community as a means of facilitating systems comparison, it has become a de-facto standard input language of many planning systems, although it is not the only modelling language for planning. Several variants of PDDL have emerged that capture planning problems of different natures and complexities, with a focus on deterministic problems.

The purpose of this book is two-fold. First, we present a unified and current account of PDDL, covering the subsets of PDDL that express discrete, numeric, temporal, and hybrid planning. Second, we want to introduce readers to the art of modelling planning problems in this language, through educational examples that demonstrate how PDDL is used to model realistic planning problems. The book is intended for advanced students and researchers in AI who want to dive into the mechanics of AI planning, as well as those who want to be able to use AI planning systems without an in-depth explanation of the algorithms and implementation techniques they use.

KEYWORDS

AI planning, problem modelling, classical planning, numeric planning, temporal planning, hybrid planning

Till mina nära fjärran
– Patrik

A mis padres, mi hermana y toda la familia, por su apoyo y enseñanzas
para disfrutar de la vida,
a Angela, por el amor, y por ser mi compañera de viaje,
y a los amigos, por tantas buenas tardes de conversaciones y risas.
– Nir

To Maria Fox and Derek Long,
for their invaluable contributions to the planning community,
and for allowing me to stand on their shoulders.
– Daniele

To K & B
– Christian

Contents

Preface

The Planning Domain Definition Language (PDDL) has been a de-facto standard modelling language in the automated planning research community for just over two decades now. In that time, the language has undergone several revisions and extensions. The specification of the syntax and semantics of different subsets of PDDL is spread over several publications and not easily accessible.

Although PDDL is intended as a "high-level" modelling language, the art of modelling planning problems in PDDL is also not easy to master. From its long use, there is a large body of tacit knowledge, such as idioms for modelling common situations, "tricks" to deal with the limitations of the language, and traps to avoid, which are also only infrequently documented.

This book presents a unified account of PDDL, as it is today, along with educational examples of how it is used to model realistic planning problems. We aim to provide an easily accessible introduction that can serve both students/researchers interested in getting into the mechanics of AI planning, as well as those whose interest is being able to use AI planning systems without needing to know too much about how they work. Explanations of some of the algorithmic and implementation techniques used in planning systems are not in the scope of this book, and can be found in general and specialised AI textbooks [Geffner and Bonet, 2013, Ghallab et al., 2004, Russel and Norvig, 2011, for example].

We would like to thank Michael Morgan at Morgan & Claypool for his great patience and constant support while writing the book. We thank Dana Nau and Sylvie Thiebaux for very constructive reviews.

We would like to acknowledge several colleagues and researchers who gave us useful feedback on earlier drafts of this book or helped us with specific topics. They include Michael Cashmore, Malte Helmert, Doug Lange, Fabio Mercorio, and Parisa Zethabi.

Patrik Haslum, Nir Lipovetzky, Daniele Magazzeni, and Christian Muise
March 2019

CHAPTER 1

Introduction

1.1 WHAT IS AI PLANNING?

Making plans is considered a sign of intelligence, and for this reason automated planning has been one of the goals of research in Artificial Intelligence (AI) since its beginning. With this goal, AI planning can be described as the study of computational models and methods of creating, analysing, managing, and executing plans.

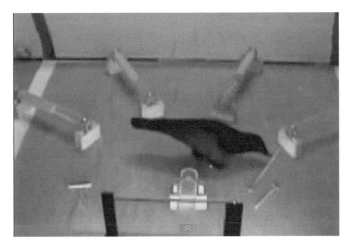

Figure 1.1: The New Caledonian crow has been observed, both in laboratory experiments and in the wild, to not only use tools to get food, but also to find and use tools to acquire better tools, and use those to get food. This could be considered a sign of planning. However, researchers are cautious about ascribing human-like cognitive processes to animals; other experiments have shown the crow failing to solve problems when it cannot directly observe the reward getting nearer. Image from Wimpenny et al. [2009].

From a more practical perspective, planning brings together AI knowledge representation and AI problem solving techniques. To apply a particular AI method (for example, heuristic search) to a problem, we must formulate a problem representation, and, most likely, devise an informative heuristic if we want the solver to be efficient. AI Planning places a third element between the problem to be solved and the solving method: a problem description language. This is a formal representation language (deriving some features from logic but adapted to express state transformation problems) in which we model the problem (see Figure 1.2).

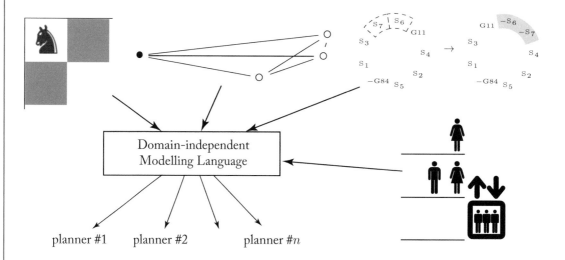

Figure 1.2: The idea of domain-independence in AI planning is to place a modelling language between the planning problem to be solved and the planning system that solves it. A problem, once modelled, can be solved (or at least we can try to solve it) with many different planners, and a planner can be applied to any problem that can be expressed in the modelling language.

An AI planning system ("planner", for short) takes the problem formalisation, or *model*, as input and uses some problem solving technique, such as heuristic search, propositional satisfiability, or other, to work out its solution. Turning the model into a search space or logical reasoning problem, creating heuristics to solve it efficiently, and so on, are problems for the designer of the planning system. The planner does not know, or need to know, what the formal problem description is about. It can be applied to any problem that can be expressed in the modelling language, though of course, not every planner will be able to solve every problem that can be given to it. This property of planners is known as *domain-independence*.

The *Planning Domain Definition Language* (PDDL) is one such problem description language. The syntax, semantics, and practical use of PDDL are the topic of this book. PDDL is not the only modelling language for planning problems (we briefly review other languages in Section 1.4.1), but owing to its role in the International Planning Competition (IPC), it is one of the most widely supported languages by planning systems. As we demonstrate throughout this book, it is also a language that has been extended in many different directions, to capture different kinds of planning problems.

The ability to plan hinges on two other capabilities: the ability to predict the consequences that the actions that can be taken will have in the world, and the ability to find, efficiently, in the space of all possible courses of action, the one that will lead to a good result. AI planning is very much focused on the second of these problems, that of finding a plan. The formalised problem description provides a predictive model from which the possible actions and the results of taking them can be deduced.

1.2   PLANNING MODELS

The plan creation aspect of AI planning deals mainly with *state transformation problems*. These are problems characterised by an initial state, a desired goal, and a set of possible actions that can be taken to change the state. A plan consists of actions that can be taken from the initial state, and which when taken will transform it into a state where the goal has been achieved[1]. Depending on the specific nature of the planning problem, the plan may be simply an ordered sequence of actions, or a schedule where actions are executed at particular times.

By making different assumptions about the model of actions, and its relation to the problem that is modelled, we arrive at different kinds of planning problems, with different difficulties. Different fragments of PDDL support the expression of several classes of planning problems.

- The simplest form of planning problem normally considered is the so-called "classical planning problem", which assumes a deterministic, discrete, and essentially non-temporal world model. We describe this fragment of PDDL in Chapters 2 and 3.

- Numeric planning relaxes the assumption of a discrete and finite model, allowing for the modelling of resources, general counting, positions in space, and many other things. We describe this fragment of PDDL in Chapter 4.

- Temporal planning extends the classical planning problem with scheduling planned actions in time. This introduces problems such as action (non-)concurrency, coordination, and scheduling actions in the presence of predictable events (for example, sunrise and sunset). We describe this fragment of PDDL in Chapter 5.

- Hybrid-system planning includes the problems above and in addition handles models with discrete modes and continuous change over time, by introducing continuous processes and exogenous events that can be triggered by plan actions or by changes in the environment. We describe this fragment of PDDL in Chapter 6.

1.3   EXAMPLES

1.3.1   THE KNIGHT'S TOUR

The knight's tour is a classic chess puzzle. In chess, the knight piece moves two spaces in any cardinal direction (horizontally or vertically) followed by one space in a perpendicular direction.

---

[1]As we will see in the following sections, planning does not always imply that the plan will be executed to bring about the goal condition. Many planning problems are hypothetical or counterfactual, in the sense that we seek a plan that could be, or could have been, performed by another actor.

Unlike other chess pieces, the knight "jumps" to the target position without stepping on intermediate squares. A knight's tour is a sequence of knight's moves that causes the piece to visit every square of a chess board, without stepping on any square twice. (Thus, it a special case of the Hamiltonian path problem.) An example of a tour, with the knight starting in the top-left corner of the chess board, is shown in Figure 1.3.

Puzzles like this are of course not the primary application of domain-independent planning, but they are useful illustrations of a certain class of planning problems[2]. The problem state can be represented as a set of facts (where the knight piece is, and which squares it has already visited) and the plan consists of a sequence of actions (the knight's moves) which transforms the starting state (the piece in its starting position, and no other square visited) to a final state satisfying the goal condition (every square visited). The actions must be formulated so that they capture all constraints of the problem: in particular, each action must not only correspond to a valid knight's move, but also enforce that the target square of the move is not already visited.

We will show how this problem can be modelled in PDDL in Chapter 2.

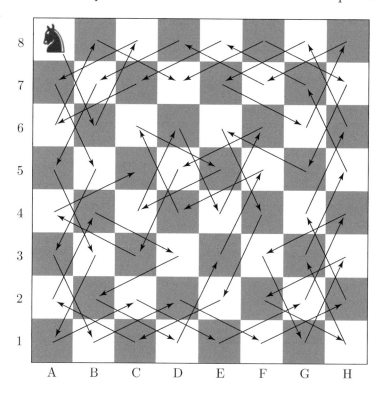

Figure 1.3: A knight's tour. The tour begins in A8 and ends in C5.

[2]This is the class of discrete and deterministic problems, which we will address in Chapters 2 and 3.

1.3.2 LOGISTICS

Planning problems arise from common operational decision problems in many business sectors. For example, managing logistics involves making decisions about when, where, and how to move and store goods. Whether it is the movement of conveyors in an automated greenhouse [Helmert and Lasinger, 2010], routing of shipments by a fleet of trucks, trains, and ships [Flórez et al., 2011], or serving passengers with a bank of elevators [Koehler and Ottiger, 2002], these can often be viewed and expressed as planning problems.

The vehicle routing problem (VRP) is a classic optimisation problem, which involves routing vehicles delivering goods to customers from a central depot so as to minimise the total distance driven [Dantzig and Ramser, 1959]. The VRP and many of its variations have a long history of study in operations research, leading to many highly efficient specialised algorithms. It is unlikely that the current generation of AI planning systems would be the best performing for this problem. However, most practical logistics problems feature additional complexities, such as time windows for deliveries, compatibility constraints between goods and vehicles (for example, refrigerated goods must be transported in a refidgerated truck), or between different types of goods (food cannot be mixed with hazardous chemicals), delivery in parts, or pickups as well as deliveries, different modes of transport (road, rail and ship, for example), and so on. A recent survey indicates that both research and commercial VRP software have yet to address many of these complications [Drexl, 2012].

Planning modelling languages, like PDDL, have the flexibility to express both the basic VRP and a wide range of extensions. We will look at modelling one particular VRP, investigated by Kilby et al. [2015], in Chapter 2.

1.3.3 PLANS AS EXPLANATIONS

Plans are not always intended for execution. In some applications of planning, the generated plan is hypothetical, demonstrating a sequence of actions that *may* have happened, or counterfactual, showing what actions could have happened if the starting state or goal had been different. Such hypothetical planning occurs in diagnostic reasoning, where the given information is a time-ordered set of observations of events that have occurred and a system model that defines which (observable and unobservable) events can occur. The task in this setting is to find one or more histories that explain the observations. Examples include alarm processing in technical networks [Haslum and Grastien, 2011], verifying and restoring consistency rules in a database [Boselli et al., 2014], and checking conformance with business process descriptions [De Giacomo et al., 2016]. The planning model typically includes some notion of "fault" events (for example, an incorrect entry made in a database), which must be hypothesised to have occurred when the observations are not consistent with the model of what should happen; the planning objective is to minimise the number, or aggregate likelihood, of the assumed faults occurring.

Another use of hypothetical plans is to measure their complexity. For example, cyber attack planning is a means of estimating how vulnerable a computer network is to cyber attacks, by planning from the perspective of a hypothetical attacker [Boddy et al., 2005, Hoffmann, 2015, Lucangeli Obes et al., 2010]. In phylogenetic analysis, the evolutionary distance between organisms can be estimated by the (weighted) shortest sequence of mutation events that can transform the genome of a hypothetical common ancestor into that of the organisms seen today. The problem of minimising such sequences of events can be formulated as a planning problem [Erdem and Tillier, 2005, Haslum, 2011].

1.3.4   PLAN-BASED CONTROL

Traditional control systems are reactive, though in many situations more efficient control can be achieved by optimising over a longer horizon. Planning can be seen as an extrapolation of receding horizon control, where a plan of control actions is computed to a goal state. Like in receding horizon control, uncertainty is dealt with by replanning. Plan-based control has been applied to realise the smart grid vision of power systems, solving problems such as the Unit Commitment problem, which is the problem of deciding which generators to switch on/off and when so as to meet forecasted demand [Piacentini et al., 2016], and voltage control [Piacentini et al., 2015]. On the energy consumer side, foresighted control of heating, ventilation, and air-conditioning (HVAC) systems in large commercial buildings can reduce energy usage by exploiting strategies such as cooling ahead of use and separating control of zones within a building, particularly when integrated with scheduling of the building's use, such as meetings or classes [Lim et al., 2015].

Control in many industrial or infrastructure systems is centred around flows, such as the flow of power in the electricity grid. Other examples include planning operations in chemical process industry [Aylett et al., 1998], and control of traffic flows via the traffic lights at intersections [Vallati et al., 2016].

In these applications, temporal and numeric planning, as discussed in Chapters 4, 5, and 6, is often required to express the dynamics of the system under control. The planners used have often been integrated with special-purpose solvers [e.g., Aylett et al., 1998, Piacentini et al., 2015], but as the capability of general-purpose planners to efficiently solve problems in more expressive fragments of PDDL improves, more and more such applications can be formulated purely in PDDL. For example, the traffic control problem modelled by Vallati et al. [2016] is formulated using processes in PDDL+. Management of multiple batteries, in which plan-based control is used decide which battery to use to serve the current load, is another example where processes are used to model the flow of charge [Fox et al., 2012].

1.3.5   PLANNING IN ROBOTICS

Planning the actions of a mobile robot is an appealing problem that has inspired AI planning research since its beginning. (The STRIPS planning system, which pioneered the precondition-effect model of classical planning, was intended for use on the robot Shakey; cf. Fikes and Nilsson [1971].) The problem of planning physically feasible, collision-free movements of a robot, whether it is a Mars rover or a multi-jointed arm, where the actions are low-level actuator controls, such as the speed of turning wheels or the degrees of joint movement, is known as *motion planning* and is well studied in robotics and AI [e.g., LaValle, 2006]. Motion planning uses models which are typically focused on geometry and kinematic constraints, and typically different from those naturally expressed in PDDL. (It is possible to express some forms of motion planning in the numeric and hybrid fragments of PDDL. We will see a simple example in Chapter 4.) However, the two forms of planning share many fundamental algorithmic ideas, and integrating motion planning with the more abstract, task-oriented planning usually represented in PDDL has been an active area of research [e.g., Cambon et al., 2009, Lagriffoul et al., 2012, Plaku and Hager, 2010, Srivastava et al., 2014, Toussaint, 2015]. Moreover, planning in a robotics context must often take account of uncertainty, due to the robot's noisy sensors and imprecise actuators. Nevertheless, mission- or task-level AI planning has found applications in planning for industrial robots [Asai and Fukunaga, 2014, Crosby et al., 2017], autonomous aircraft [Bernardini et al., 2017a] and underwater vehicles [Cashmore et al., 2014, McGann et al., 2008], space missions [Ai-Chang et al., 2004, Jonsson et al., 2000], and more. The ROSPlan library [Cashmore et al., 2015] allows any PDDL-capable planner to be integrated in the ROS framework. It handles invoking the planner, and plan dispatch and monitoring, and includes interfaces to several common ROS libraries.

1.3.6 NARRATIVE PLANNING

The plot of a story has some similarity with a plan, in that they are both sequences of events that change the state of the (story) world. For a story to be coherent and plausible, the event sequence must be logically possible and connected by causes and effects. This has inspired a long line of AI research into planning-based approaches to automating the creation or presentation of stories (cf. Gervás [2009] for a survey of early work, and Brenner [2010], Ware and Young [2014], and Porteous et al. [2013] for more recent examples).

Planning-based story generation raises many challenges: most actions in a story are taken by characters of the story, who, in order to be believable, must be seen as acting in a manner that is directed by their own goals and desires. Yet characters' goals are often in conflict, and in conflict with the goals of the author. For example, in a Shakespearian tragedy it is often the author's goal that all the main characters die, but that is certainly not the intention of the characters, who fight against their fate. If instead they went helpfully to their graves in Act 1, the story would be both boring and unbelievable. This has been addressed by modifying planners to consider character goals [e.g., Ware and Young, 2014], multi-agent planning and plan simulation [e.g., Brenner, 2010], or re-formulating the problem to make character's goals and reasoning explicit [e.g., Chang and Soo, 2008, Haslum, 2012].

Another difficulty is the scale of the problem. To permit the planner to come up with varied stories, so that its creative potential is not unnecessarily restricted, the planning model must include a sufficient breadth of events that could occur in the story world. Creating this model by hand places an impractical burden on system designers, driving investigations into methods to automatically or semi-automatically build or extend planning models [see, e.g., Chambers and Jurafsky, 2009, Porteous et al., 2015, Sil and Yates, 2011].

Finally, plot creation differs from most other applications of planning in that it is unclear how to measure the quality of a candidate plan. While in most other cases the objective is to minimise plan cost or complexity, a story must have a certain amount of complexity to make it interesting. A different way to approach the difficulty of judging what is a good story plan is to generate a collection of plans that is *diverse* and leave humans to make the final choice (see, for example, Srivastava et al. [2007] or Goldman and Kuter [2015] for discussion of measures of plan diversity and how to generate diverse plans).

1.4   THE ORIGINS OF PDDL AND THE SCOPE OF THIS BOOK

Working out a course of action to achieve a goal is one of the earliest applications of logical reasoning in AI [e.g., Green, 1969]. The STRIPS planner ["*STanford Research Institute Problem Solver*", Fikes and Nilsson, 1971] is often credited as the first system to implement a dedicated algorithm for planning, and although this algorithm has long since been superseded, its approach to modelling actions, with preconditions, positive and negative effects expressed as sets of atomic facts, survives to this day. (The simplest subset of PDDL, which we present in Chapter 2, is commonly referred to as "STRIPS".)

PDDL was created as a common language for the first International Planning systems Competition (IPC) in 1998 [McDermott, 2000]. PDDL version 1 was defined by a committee of researchers in the field of AI planning, led by Drew McDermott [McDermott et al., 1998]. New versions of PDDL have been proposed in conjunction with later editions of the planning systems competition. The subsequent versions have expanded the language with new features to express kinds of planning problems of different kinds (e.g., numeric and temporal planning), with the aim of directing researchers to develop planning systems capable of handling these problem classes. Over time, however, language features that failed to gain popularity have also been removed. The PDDL versions published so far are as follows.

- PDDL 1 [McDermott et al., 1998].

- PDDL 1.2 [Bacchus, 2000]. This version trimmed some unused features from the language, and corresponds to most of the classical planning formalism presented in Chapters 2–3.

- PDDL 2.1 [Fox and Long, 2003]. This version extended the language to represent numeric planning (presented in Chapter 4) and temporal planning problems (presented in Chapter 5). Exploiting the expressivity of the numeric planning extension, it also introduced a syntax for specifying an objective function for optimisation.

- PDDL+ [Fox and Long, 2006]. This version further extended the numeric and temporal representation to hybrid planning (presented in Chapter 6).

- PDDL 2.2 [Edelkamp and Hoffmann, 2004]. This version added two features: Axioms (described in Chapter 3), which add more expressive conditions to classical planning, and timed initial literals, which provide a syntactic convenience for defining a schedule of predictable events in temporal planning.

- PDDL 3.0 [Gerevini et al., 2009]. This version added syntax for temporally extended goals and preferences to classical planning. This is covered in Chapter 3.

- PDDL 3.1 [Helmert et al., 2008]. This version defined the restricted syntax for specifying action costs (described in Chapter 2). It also introduced general finite-domain state variables, known as "object fluents". We do not cover this extension in this text.

Although PDDL is intended to be a common modelling language, for a certain class of planning problems, it is important to recognise that it is not a standard. It is better understood as a community effort to promote interchangeability and ease of application of planning systems. Discrepancies between the different versions, and differing styles and levels of formality in the articles and technical reports cited above, mean that there is no complete and unambiguous specification of the syntax or semantics of all of PDDL. Moreover, planners usually support only a subset of (some version of) PDDL, and different systems' implementors have interpreted the ambiguous aspects of the language in different ways.

Yet, there is a core of PDDL about which there is a reasonably broad consensus of understanding. Our approach in this book is to focus on that core. We cover elements from all PDDL versions, but have omitted some of the unclear, unused, or as yet underdeveloped parts of the language from each. As much as practical, we also point out specific areas where we are aware that different planners may interpret the language differently.

1.4.1 OTHER PLANNING LANGUAGES

Although PDDL is the modelling language used by many planners, it is far from the only planning language. What defines domain-independent planning is the idea of using some declarative problem definition formalism, not what that formalism is. Many different classes of planning problems can be distinguished, and while different fragments of PDDL capture some of them, there are others that do not fit into PDDL at all. This book is about PDDL, and planning problems that can be expressed in PDDL. In this section we briefly review some alternative planning languages, and the kinds of problems that they are intended for.

Extensions of PDDL have been defined to model problems with uncertainty, in the form of a partially known initial state and actions with nondeterministic effects. This uncertainty may be quantified with probabilities [Younes and Littman, 2004] or express only a set of possibilities. Planning, when viewed as a state transformation problem, with probabilistic uncertainty closely resembles solving an MDP or POMDP [Mausam and Kolobov, 2012, Puterman, 1994]. RDDL [Sanner, 2011] is a language designed for probabilistic planning. While it's syntax is quite different from PDDL, it is based on the same state-transition model. Another example that extends PDDL is MA-PDDL [Kovacs, 2012], which is designed to model multi-agent planning problems for planning both *for* and *by* several agents. We describe some of the languages most closely related to PDDL in greater detail later in Chapter 7.

The timeline-based planning model [e.g., Ghallab and Laruelle, 1994, Muscettola, 1994] was developed for temporal planning problems with a significant scheduling component. Fundamental to this model is a timeline, which represents the evolution of a state variable over time. State persistence and state change are treated uniformly, and actions are modelled mainly through sets of constraints that they impose across timelines. Initial conditions and goals are also expressed as constraints over timelines, both possibly referring to several points in time. Languages for modelling problems of this kind are mostly specific to planning systems [e.g., Barreiro et al., 2012, Chien et al., 2000, Frank and Jonsson, 2003]. ANML [Smith et al., 2008] is a recent effort to create a unifying modelling language for timeline-based planning.

Hierarchical task network (HTN) planning embodies a different view of what planning is. Instead of changing the state of the world model, a HTN planning problem is defined as a set of abstract *tasks* to be done, and set of methods for each task that represent different ways in which it can be carried out (although methods may also have conditions and effects on state). A plan recursively decomposes the given abstract tasks into primitive tasks, which are concrete, executable actions. There is no widely shared modelling language for HTN planning, as different HTN planners use their own input languages [e.g., Nau et al., 2005, Tate et al., 1994, Wilkins, 1986]. However, ANML also includes some hierarchical planning features.

Related to both timeline-based and hierarchical planning, languages like RMPL [Effinger et al., 2009, Williams and Ingham, 2002] and Golog [Levesque et al., 1997] express planning problems as nondeterministic programs. They can use programming constructs, such as conditional branching or iteration, as well as temporal constraints, to express a high-level strategy. The task of the planning system becomes to instantiate the unspecified choices in the program to achieve a successful execution from a given initial situation.

1.4.2 RELATION TO NON-PLANNING FORMALISMS AND OTHER FIELDS

AI planning is not the only area of computer science that defines a class of problems via a modelling framework and aims to develop general solvers for the class. Propositional satisfiability (SAT) [Biere et al., 2009], its extension to quantified boolean formulas (QBF) [Nethercote et al., 2007, Schaefer, 1978], (finite-domain) constraint programming (CP) [Rossi et al., 2006], and mathematical programming formalisms such as linear programming (LP) and mixed-integer programming (MIP) [Dantzig, 1960, Wolsey, 1998] are prominent examples. The emphasis on the modelling language varies: for example, there are many different formats for specifying a MIP, yet there is a common, unambiguous definition of what is the class of MIP problems. Likewise, the emphasis on domain-independence, in the sense of fully automatic generality, of solvers varies. SAT solvers, like AI planners, are typically intended to work on any input without problem-specific configuration; in CP, and to some degree MIP, on the other hand, the modelling of a problem and customisation of the solver often go hand in hand.

In terms of the complexity of problems that can be expressed, AI planning, even when restricted to only the classical subset of PDDL, sits in a higher complexity class than many other general-purpose formalisms (PSPACE-complete for grounded PDDL, EXPSPACE-complete for parameterised PDDL). Some of the more expressive fragments of PDDL are undecidable in general. In contrast, SAT, CP, and MIP are all in NP, and thus comparatively "easy". QBF is also PSPACE-complete, yet very different in the types of problems it naturally models. Many AI planners are built on heuristics, and often are able to find some solution (plan) for a complex problem quickly, but offer no guarantee of optimality, or even completeness. Complete and optimal planners typically target subclasses of planning problems, such as classical planning with additive action costs. In spite of the difference in worst-case complexity, reductions of planning into other formalisms have been used effectively [Kautz and Selman, 1996, van Beek and Chen, 1999]. Fixing a low polynomial bound on plan length (which in theory sacrifices completeness, but often works in practice) allows, for example, classical planning to be reduced to SAT or CP, and numeric planning to be reduced to MIP.

Another computer science area that is closely related is model checking [Clarke et al., 1993], which is concerned with verifying (or disproving) properties of reactive, usually discrete-state, systems, such as for example communications protocols. These properties, usually expressed in temporal logics such as LTL, correspond to the existence or non-existence of execution traces, i.e., sequences of transitions, in the system. There is a clear analogy with planning, as a plan is also a sequence of transitions in the state space of the planning problem. Reductions of both classical planning and other planning variants to model checking have been shown [e.g., Giunchiglia and Traverso, 1999], as well as translations in the other direction, i.e., stating a verification problem as a planning problem [Camacho et al., 2018, Ghosh et al., 2015, Patrizi et al., 2013]. We will show an example, encoding the famous "dining philosophers" deadlock detection problem in PDDL, in Section 2.4.2.

However, there is a clear difference in emphasis: in planning, the emphasis is on efficiently finding a plan, and in some cases on finding a plan of good quality, while in model checking the emphasis is on proving that the specification is satisfied, which in many cases translates to proving that no plan exists. The problem of proving plan non-existence has received increased attention in planning research recently [see, e.g., Muise and Lipovetzky, 2016], and many of the techniques developed in the field of model checking, which are useful not only for proving plan non-existence but also in proving plan optimality, have been adapted into planners.

1.5   PLANNING SYSTEMS AND MODELLING TOOLS

The main purpose of modelling planning problems in PDDL is to apply automated planning systems to find solution plans. However, there are also other tools that can be valuable in the process of writing (and debugging) a PDDL model.

`http://planning.domains` is an on-line repository of planning benchmark models, which also includes an on-line editor with PDDL-specific features such as syntax highlighting and semi-automatic instantiation of some common model patterns. Throughout the book, we provide links that preload examples into this editor. It also provides an interface to a planner as a web service, and from within the editor. However, the planner it makes available does not support all of PDDL, and as it has a runtime limit of 15 s it is restricted to solving small problems.

For readers who want to try other planners, we can suggest exploring the archive of planning competition web pages at `http://icaps-conference.org/index.php/Main/Competitions` (the "deterministic" tracks). Each of the competitions since 2008 has provided links to the source code of participating planning systems. However, these are "snapshots" of the planners at the time of the competition, and many of them are not maintained.

The VAL tool suite (`https://github.com/KCL-Planning/VAL`) includes a PDDL syntax checker and a plan validator. A plan validator is a tool that takes as input a problem definition (in PDDL) and a plan, and determines if the plan solves the problem. Validating manually written plans can be a useful approach to debug the problem definition. An alternative implementation of a plan validator for PDDL is INVAL (`https://github.com/patrikhaslum/INVAL`).

As we have mentioned, PDDL is not a standard, and there are some aspects of the language where its specification is vague, ambiguous, or even missing. Moreover, planners and plan validators are research prototypes. They usually support only a subset of PDDL, their interpretation of some aspects of the language may differ, and, like any complex software, they may have bugs. Even the plan validators occasionally disagree on whether a problem definition or a plan is valid. When modelling a problem in PDDL, one must always be prepared to reformulate the model to work around planners' limitations and quirks.

CHAPTER 2

Discrete and Deterministic Planning

PDDL supports the expression of several different classes of planning problems. In this chapter and the next, we begin with the simplest class of planning problems, which are discrete, finite, and deterministic. This kind of planning is often called *classical* (to distinguish it from the many different extensions that have later followed).

In PDDL, the state of the world we model is described by the set of *facts* that are true in it. The actions in a plan cause state transformations, which alter the set of true facts according to the action's *effects*. Further, each action has a *precondition*, which determines in which states the action can be applied. The classical planning problem makes the following assumptions.

- The set of facts that may be true or false (or, in other words, the set of state variables) and the set of possible actions is finite.

- The models of actions available to the planner are correct and deterministic. This means that applying the same action in the same state always leads to the same result.

- The planner has complete knowledge of the initial state (there are no unknown facts), and the world is static, meaning that there are no changes to the state other than those caused by the planned actions.

These may appear to be highly unrealistic and limiting assumptions. Nevertheless, the classical planning model is useful for several reasons. First, since it provides a simplified, idealised setting, it is a good starting point for studying and developing automated planners. Many of the lessons learned and techniques discovered for classical planning have generalised to, or inspired adaptations of methods for, more complex planning problems.

Second, in many cases even the classical planning model is enough to express a sufficiently good approximation of a problem. The formal model of a planning problem is usually an abstraction of the real problem. Aspects of a problem that we know can be resolved when the plan is executed do not need to be considered at planning time. For example, if we plan to walk from one place to another, we do not plan ahead in detail every leg movement or placement of feet on the ground. The plan executor—whether it is a robot, control system, or human—has some known capabilities (or "skills"), and the actions of the planning model are what it can do using those skills. The same is true of surprises, whether they arise from incomplete knowledge of the state or from an incomplete model of the preconditions or effects of actions. In any realistic situation, the list of unexpected things that could happen—actions failing or having unexpected side-effects, or another actor interfering with the state while planning is done—is endless. But if we cannot make plans to avoid unexpected disruptions, nor plan in advance how to recover from them if they happen, then it is not useful to include them in the planning model.

In the next section we will introduce the basic building blocks for PDDL by way of a small number of examples of increasing complexity. We follow in Section 2.2 with a formal discussion of *plans*—the desired solution of a planning problem—and then discuss some further details of the syntax of the classical planning subset of PDDL in Section 2.3. We detail some more advanced modelling examples in Section 2.4, and finish with a discussion on the complexity of solving planning problems in Section 2.5.

2.1   DOMAIN AND PROBLEM DEFINITION

PDDL divides the definition of a planning problem into two parts: the *domain* defines the state variables (facts that may be true or false) and actions, while the *problem* defines initial state and the goal condition. Thus, a domain definition is a general model of the relevant aspects of the world in which we are planning, while the problem definition is a specific problem instance in this domain that specifies where we begin and what we must achieve. The domain and problem definitions are usually placed in two files, using the extension ".pddl". This is not mandated by the PDDL language, but many planners require it.

In the remainder of this section, we introduce the basic features of PDDL through a series of annotated examples.

2.1.1   PDDL: FIRST EXAMPLE

We begin with a very simple example (a kind of planning "hello world"). Imagine that we have a switch: the switch can be in one of two positions ("on" or "off") and the two actions we can take is to move it from one position to the other. Here is a complete definition of this domain in PDDL:

```
(define (domain switch)
  (:requirements :strips)

  (:predicates (switch_is_on)
               (switch_is_off))

  (:action switch_on
   :precondition (switch_is_off)
   :effect (and (switch_is_on)
```

```
                (not (switch_is_off)))
   )

  (:action switch_off
   :precondition (switch_is_on)
   :effect (and (switch_is_off)
                (not (switch_is_on)))
   )
)
```

PDDL Example 1: A domain definition for the switch example.

We can make several observations about the language from this example: All PDDL expressions are enclosed in matching parentheses. The order of expressions within a pair of parentheses is prefix, meaning operator followed by arguments. For example, the conjunction that appears in the effects of the actions is written (`and` `<conjunct1>` ... `<conjunctN>`). These syntactical conventions are inherited from the LISP programming language.

Most keywords start with a colon (`:`), but there are some exceptions (for example, `define`, `domain`, and the logical operators `and` and `not`). The remaining terms in this example are names (the name of the domain, and the names of predicates and actions). The PDDL version 1 specification [McDermott et al., 1998] states that valid names "are strings of characters beginning with a letter and containing letters, digits, hyphens (-) and underscores (_). Case is not significant". However, many planners will only accept a narrower set. Names made up of alphanumeric characters (a–z, 0–9) and underscores (_) should be acceptable to all planners. Note that names are case insensitive. Line breaks have no semantic meaning in PDDL. Whitespace in general is ignored, except that it is required to separate names, keywords, and other lexical elements where they appear in sequence without an intervening parenthesis.

As we have mentioned, different subsets of PDDL allow the specification of different classes of planning problems. The `:requirements`, section of the domain definition indicates which features of PDDL the domain uses, and thus, to some extent, what kind of planning problem it is. In this example, the keyword `:strips`, indicates that the domain is of the simplest form. The requirements specification is somewhat redundant—most planners will examine the domain definition itself to determine if it uses any PDDL features that they cannot deal with rather than rely on the requirements keywords. It is, however, good practice to always write a correct requirements specification.

The `:predicates` section of the domain definition contains the list of the model's state variables. These are binary variables, meaning they represent facts that are either true or false. In this small example, there are only two state variables and therefore only four possible states, which are shown in Figure 2.1. Actions define transitions between states. Each action has a precondition, which defines when the action is applicable, and an effect that defines what happens when the action is applied. In the :strips fragment, an action's precondition can be a single fact or a conjunction of facts. For example, in Figure 2.1, action (switch_on) is applicable only in states where the fact (switch_is_off) is true (the two states on the right). The effect is written as a conjunction of facts made true and facts made false. The latter are written as negated literals. Thus, the effect of, for example, the switch_on action is to make (switch_is_on) true and (switch_is_off) false. Note that if this action is applied to the state where both facts are already true (lower right in Figure 2.1) only one of them changes.

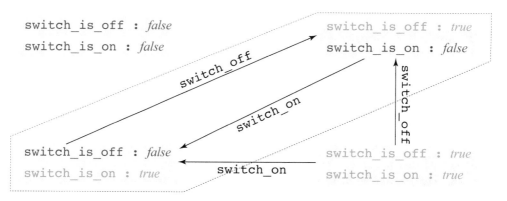

Figure 2.1: All possible world states of the model in the switch domain, Example 1, and the transitions between states caused by actions. The encircled states are the intended reachable part of the state space, i.e., those that correspond to actual possible states of the switch.

In this example, it is of course our intention that at any time exactly one of the facts (switch_is_on) and (switch_is_off) should be true, and the other false, since the switch can only be in one position. That is, of the four states in Figure 2.1 only two are "valid" (these two states are encircled with a dotted line). PDDL has no general mechanism for enforcing this kind of constraint on states (although it does allow for negative literals in action preconditions, as we will see later in this section, which allows us to eliminate these "spurious" states in this example with a modified model). However, the way in which we have written the two actions ensure that if we start from one of the states in the valid set, no action can take us out of it. If the initial state lies in this set, we refer to it as the *reachable* state space. A property that holds in all reachable states is often called a *state invariant*. Our task when defining a planning domain is to write it so that the reachable state space corresponds to that of our intended problem.

The PDDL problem definition specifies the initial situation and the goal condition. (There are more components to planning problems that will be introduced later in this section.) Here is an example problem definition for our simple switch domain.

```
(define (problem turn_it_off)
  (:domain switch)

  (:init
   (switch_is_on)
   )

  (:goal (switch_is_off))
)
```

PDDL Example 2: A problem definition for the switch example.

Some key observations we can make from this are the following. The problem, like the domain, is named. The problem definition also includes a reference to the domain that it is associated with, in the form (:domain *<name>*). Note that, much like the :requirements specification, this information is somewhat redundant: most planners require as input two files, one with a domain definition and the other with a problem definition, and assume the two go together. It is, however, good practice to name the right domain in the problem definition.

The (:init ...) section defines the initial state of the problem instance, by listing all facts that are true in that state. Every fact that is not explicitly mentioned is assumed to be false initially. This means that in this example, the initial state is the lower left state in Figure 2.1. Note that the :init keyword is followed directly by a list of facts; there is no and, or additional parentheses.

The :goal section specifies a condition that must be satisfied at the end of a valid plan. The goal condition has the same form as an action precondition; for :strips, the form is either a single fact or a conjunction of facts. In this example, the goal is simply for the switch to be off. Note that the goal condition does not have to specify a unique state, or even a unique reachable state. As we shall soon see, however, more expressive planning formalisms relax the restriction on how we express a goal.

2.1.2   EXAMPLE: MODELLING THE KNIGHT'S TOUR

In this section we show how to model the Knight's Tour problem, described in Section 1.3.1. We will show how using *parameters* and *objects* we can write a PDDL domain and problem that is more compact and general. But first, let's look at how to formulate the Knight's move as an action.

Tool support: PDDL syntax checking

The VAL tool suite, available at `https://github.com/KCL-Planning/VAL`, includes a PDDL syntax checker, called `parser`. It is simple to use:

`$ parser domain-file.pddl problem-file.pddl`

The checker's output can be quite verbose, because it issues warnings for every item in the PDDL definition that implies a missing `:requirements` keyword.

Another useful tool is `https://planning.domains`. This is an on-line repository of PDDL domain and problem definitions, but includes also an editor with PDDL syntax highlighting and an option to run a planner.

```
(:action move_A8_to_B6
 :precondition (and (at_A8)
                    (not (visited_B6)))
 :effect (and (not (at_A8))
              (at_B6)
              (visited_B6))
 )
```

PDDL Example 3: A possible formulation of a Knight's move as an action.

The action shown above moves the Knight from one square A8 to square B6, which is a valid Knight's move. (We follow the convention of naming columns with letters and rows with numbers. Figure 1.3, on page 4, shows the naming scheme.) The action's precondition requires that the Knight is on the origin square, and that the destination square has not already been visited (recall that the aim of the puzzle is for the Knight to visit every square exactly once). Note that we use the negation of the predicate here: this is allowed in PDDL if the keyword `:negative-preconditions` is added to the `:requirements` section. The effect makes predicate `(at_A8)` false and predicate `(at_B6)` true, representing the fact that the location of the piece has changed; it also makes `(visited_B6)` true, so that the Knight won't return to this square later in the plan. There are 336 valid Knight's moves on a chess board. Thus, we could model the puzzle by writing 336 actions like this one, using 64 predicates for the possible current locations of the Knight and 64 to keep track of the squares visited. However, these actions are all very similar: they differ only by which two squares the Knight moves between.

Parameterising predicates and actions allows us to write a smaller number of predicates and actions—in the case of the Knight's tour, only one action—that are then *instantiated* with a collection of possible objects. The following is a parameterised version of the Knight's Tour domain:

```
(define (domain knights-tour)
  (:requirements :negative-preconditions)

  (:predicates
    (at ?square)
    (visited ?square)
    (valid_move ?square_from ?square_to)
  )

  (:action move
   :parameters (?from ?to)
   :precondition (and (at ?from)
                      (valid_move ?from ?to)
                      (not (visited ?to)))
   :effect (and (not (at ?from))
                (at ?to)
                (visited ?to))
  )
)
```

PDDL Example 4: A parameterised domain formulation for the Knight's Tour Puzzle.

We have included the PDDL requirement keyword `:negative-preconditions`, since the precondition of the `move` action uses a negated literal in its precondition.

The key new element in this domain is the use of parameters in the predicate and action declarations. Parameter symbols in PDDL must begin with a '?', and otherwise follow the rules for valid names. The parameters used in the declaration of a predicate must be all different. That is, we could not have written the predicate declaration (`valid_move ?square ?square`). In the action definition, the list of parameters is written in the `:parameters` field. To remove any potential ambiguity in the action's meaning, literals in the preconditions and effects of an action can only use parameter symbols that are declared as parameters of the action. Note that these do not need to equal the parameters used in the predicate's declaration. We often refer to a parameterised action definition as an *action schema*.

The values that parameters can assume correspond to objects declared in the PDDL problem definition. Here is a portion of the problem definition that goes with the domain in Example 4:

```
(define (problem knights-tour-problem-8x8)
  (:domain knights-tour)

  (:objects
    A1 A2 A3 A4 A5 A6 A7 A8
    B1 B2 B3 B4 B5 B6 B7 B8
    ...
    H1 H2 H3 H4 H5 H6 H7 H8)

  (:init
    ; The Knight's starting square is arbitrary; here, we have
    ; chosen the upper right corner.
    (at A8)
    (visited A8)

    ; We have to list all valid moves:
    (valid_move A8 B6)
    (valid_move B6 A8)
    (valid_move A8 C7)
    (valid_move C7 A8)
    (valid_move B8 A6)
    (valid_move A6 B8)
    (valid_move B8 C6)
    ...
  )

  (:goal (and (visited A1)
              (visited A2)
              ...
              (visited H8)))
)
```

PDDL Example 5: Part of the PDDL problem definition for the Knight's Tour Puzzle.

We now have an `:objects` section that lists all of the objects in the problem instance. In this example, the objects are squares of the chess board ("…" means we have abbreviated the listing for clarity). Here we also have our first example of comments in PDDL. A comment begins with a semicolon (`;`) and continues to the end of the line. (This comment syntax also derives from the fact that PDDL's syntax is based on that of LISP.) The `:init` and `:goal` sections, like before, list the facts that are true in the initial state and the conjunction of facts that must be true for the goal to be achieved, respectively. Note that they are lists of *ground* facts, meaning all predicate parameters must be instantiated with objects of the problem.

The predicate `valid_move` is an example of what is often called a *static predicate*, meaning a predicate that is not affected by any action. There is no special syntax to mark the predicate as static—it is only the fact that it does not appear in the `:effect` of any action that makes it so. Because a static predicate can not change, the instances of it that are true in the initial state are exactly those that will be true in any reachable state. Static predicates often play an important role in parameterised PDDL formulations by restricting the instantiation of action parameters. Consider the `move` action in the formulation of the Knight's Tour domain above: without (`valid_move ?from ?to`) in its precondition, this action would allow the Knight to jump between any pair of squares on the board. The presence of this literal in the precondition means that the action can be instantiated only with moves between those pairs that are explicitly listed as valid in the `:init` section of the problem. However, this also means that we have to list all 336 valid Knight's moves in the `:init` section.

A more compact formulation can be found by recognising the regularities among the valid moves. For example, for every valid move, the reverse move is also valid. Thus, we could write out only one direction for each move, and define an action `move_back` identical to move except for having the precondition (`valid_move ?to ?from`) with the two parameters reversed. But we can do better than that, by separating the two coordinates of each square into different parameters. Examples 6 and 7 below show the resulting domain and problem, respectively. They can also be found at editor.planning.domains/pddl-book/knights_tour_2.

```
(define (domain knights-tour)
  (:requirements :negative-preconditions)

  (:predicates
    (at ?col ?row)
    (visited ?col ?row)
    (diff_by_one ?x ?y)
    (diff_by_two ?x ?y)
  )

  ; Action move_2col_1row moves the Knight two steps along the horizontal
```

```
; axis (changing the column by 2) and one step along the vertical axis
; (changing the row by 1), while action move_2row_1col below does the
; opposite.

(:action move_2col_1row
 :parameters (?from_col ?from_row ?to_col ?to_row)
 :precondition (and (at ?from_col ?from_row)
                    (diff_by_two ?from_col ?to_col) ; col +/- 2
                    (diff_by_one ?from_row ?to_row) ; row +/- 1
                    (not (visited ?to_col ?to_row)))
 :effect (and (not (at ?from_col ?from_row))
              (at ?to_col ?to_row)
              (visited ?to_col ?to_row))
 )

(:action move_2row_1col
 :parameters (?from_col ?from_row ?to_col ?to_row)
 :precondition (and (at ?from_col ?from_row)
                    (diff_by_two ?from_row ?to_row) ; row +/- 2
                    (diff_by_one ?from_col ?to_col) ; col +/- 1
                    (not (visited ?to_col ?to_row)))
 :effect (and (not (at ?from_col ?from_row))
              (at ?to_col ?to_row)
              (visited ?to_col ?to_row))
 )

)
```

PDDL Example 6: Domain definition for an alternative formulation of the Knight's Tour Puzzle.

```
(define (problem knights-tour-problem-8x8)
  (:domain knights-tour)

  ; Define a set of "numbers" 1..8:
  (:objects n1 n2 n3 n4 n5 n6 n7 n8)

  (:init
    ; Initial position of the Knight piece (upper left corner):
```

```
(at n1 n8)
(visited n1 n8)

; Here, we have to list all instances of the static
; predicates diff_by_two and diff_by_one:
(diff_by_one n1 n2)
(diff_by_one n2 n1)
(diff_by_one n2 n3)
(diff_by_one n3 n2)
...
(diff_by_one n7 n8)
(diff_by_one n8 n7)

(diff_by_two n1 n3)
(diff_by_two n3 n1)
(diff_by_two n2 n4)
(diff_by_two n4 n2)
...
  )

(:goal (and (visited n1 n1)
            (visited n1 n2)
            ...
            (visited n8 n8)))
  )
```

PDDL Example 7: Part of the PDDL problem definition for the second formulation of the Knight's Tour Puzzle.

In this formulation, we use a pair of objects, one for row and one for column, to represent each square on the chess board. As a result, the predicates at and visited now have two parameters, and each version of the move action has four. The main simplification lies in the encoding of valid moves: because every valid Knight's move changes either the row by ±2 and and the column by ±1 or vice versa, we can encode these conditions on the two pairs of parameters (?from_row and ?to_row, and ?from_col and ?to_col, respectively) separately. As can be seen in the new problem definition (Example 7), by using the same objects for positions on both axes we now only need eight, as well as fewer instances of the static predicates. The eight objects represent integers 1–8. (They are named n1 through n8 because names in PDDL cannot consist of digits only, and many planners require names to start with an alphabetic character.)

We will see the use of objects that represent a subset of integers, along with static predicates encoding arithmetic relations over them, again in the following sections.

These examples illustrate that in PDDL, as in any modelling language, there may be many possibilities to model the same problem or concept. In the formulation that explicitly lists all valid moves it is easier to modify what they are. For example, if we wanted to try to solve the Knight's tour on a "mutilated chess board" (one from which the two white corner squares are cut out) or a "doughnut chess board" (one from which the four squares at the centre are removed), this formulation could easily be adapted by simply removing the objects corresponding to the missing squares and any predicates that mention them from the problem definition. Thus, which formulation is preferable often depends on exactly what is the set of problem instances we want to solve.

2.1.3   EXAMPLE: LOGISTICS

In this section, we will formulate a small-scale practical logistics problem, studied by Kilby et al. [2015], in PDDL. We will use this example to introduce two more features of the PDDL language: our model will have objects of different *types*, and we will assign each action a *cost*, so that we can express the objective function.

The problem is the single-day linehaul problem of a consumer goods distributor, who uses a fleet of trucks to deliver goods requested by a set of customers. The goods are so-called "fast moving consumer goods"—in other words mainly fresh food—and customers are supermarkets and other retailers. Each truck starts from a central depot, visits a sequence of customers, and returns to the depot at the end. Because the distance from the depot to the area where the customers are is significant, each truck can make only one such tour in a day. Goods are divided into chilled and ambient, and each customer has requested a certain quantity of each type. Chilled goods can only be transported in refrigerated trucks, while ambient-temperature goods can be sent both on trucks with refrigeration and without. Besides being refrigerated or not, the available truck models also have different capacities and per-kilometer driving costs. The objective is to minimise the total cost which is the sum over all trucks of their per-kilometer cost times the distance travelled. On a given day, some trucks may not be used at all. Figure 2.2 shows a small example instance of the problem.

We will also see the encoding of numbers as objects, used for positions on the chess board in the second encoding of the Knight's tour above, again in the formulation of the linehaul problem. Here, however, the numbers represent capacities and quantities of goods instead of coordinates. The numeric planning fragment of PDDL, which we introduce in Chapter 4, allows for general numeric functions and expressions. However, because customer requests and truck capacities in this problem are integers (with relatively small ranges) we can also use the encoding of these as objects.

As mentioned in Section 1.3.2, practical logistics problems often have many more complexities, such as delivery time windows, multi-modal transport, pickup-and-delivery, combined

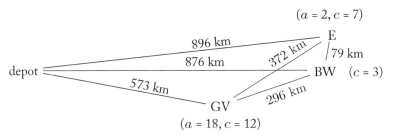

Figure 2.2: A small example instance of the Linehaul problem. The numbers next to customer locations show their demand for ambient and chilled goods.

routing and fleet selection, and so forth. Many of these can be formulated as variations of the model we present here, although some may require the use of a more expressive subset of PDDL. For example, to model time windows and other scheduling constraints we need the temporal planning subset, introduced in Chapter 5. In Chapters 3 and 4, we will introduce PDDL features that simplify modelling more complex objective functions.

Introducing Object Types
The following is a first step toward a domain definition. (The domain and example problem can be found at editor.planning.domains/pddl-book/linehaul_without_costs.) Here, we have not yet introduced action costs.

```
(define (domain linehaul_without_costs)
  (:requirements :strips :typing)

  (:types
    location truck quantity - object
    refrigerated_truck - truck
  )

  (:predicates
    (at ?t - truck ?l - location)
    (free_capacity ?t - truck ?q - quantity)
    (demand_chilled_goods ?l - location ?q - quantity)
    (demand_ambient_goods ?l - location ?q - quantity)
    (plus1 ?q1 ?q2 - quantity)
  )

  ;; The effect of the delivery action is to decrease demand at
```

```
;; ?l and free capacity of ?t by one.
(:action deliver_ambient
  :parameters (?t - truck ?l - location
               ?d ?d_less_one ?c ?c_less_one - quantity)
  :precondition (and (at ?t ?l)
                     (demand_ambient_goods ?l ?d)
                     (free_capacity ?t ?c)
                     (plus1 ?d_less_one ?d)  ; only true if x={?d,?c}, x > n0
                     (plus1 ?c_less_one ?c))  ; and x = x_less_one + 1
  :effect (and (not (demand_ambient_goods ?l ?d))
               (demand_ambient_goods ?l ?d_less_one)
               (not (free_capacity ?t ?c))
               (free_capacity ?t ?c_less_one))
)

(:action deliver_chilled
  ;; Note type restriction on ?t: it must be a refrigerated truck.
  :parameters (?t - refrigerated_truck ?l - location
               ?d ?d_less_one ?c ?c_less_one - quantity)
  :precondition (and (at ?t ?l)
                     (demand_chilled_goods ?l ?d)
                     (free_capacity ?t ?c)
                     (plus1 ?d_less_one ?d)  ; only true if x={?d,?c}, x > n0
                     (plus1 ?c_less_one ?c))  ; and x = x_less_one + 1
  :effect (and (not (demand_chilled_goods ?l ?d))
               (demand_chilled_goods ?l ?d_less_one)
               (not (free_capacity ?t ?c))
               (free_capacity ?t ?c_less_one))
)

(:action drive
  :parameters (?t - truck ?from ?to - location)
  :precondition (at ?t ?from)
  :effect (and (not (at ?t ?from))
               (at ?t ?to))
)
)
```

There are several new elements in this domain definition.

- We added the keyword :typing to the requirements, to reflect that this domain uses objects of different types.

- We have added a :types section, in which we declare the names of the object types used.

- Finally, we have added a type specification to every parameter in the predicate and action declarations.

The following is the (abbreviated) problem definition for the example instance in Figure 2.2.

```
(define (problem linehaul-example)
  (:domain linehaul_without_costs)

  (:objects
    ADoubleRef - refrigerated_truck
    BDouble - truck
    depot GV E BW - location
    ;; Declare objects for integers, up to the largest quantity
    ;; that appears in the problem:
    n0 n1 n2 ... n40 - quantity
    )

  (:init
    (at ADoubleRef depot)
    (at BDouble depot)
    (free_capacity ADoubleRef n40)
    (free_capacity BDouble n34)
    (demand_chilled_goods GV n18)
    (demand_ambient_goods GV n12)
    (demand_chilled_goods E n7)
    (demand_ambient_goods E n2)
    (demand_chilled_goods BW n3)
    (demand_ambient_goods BW n0)
    (plus1 n0 n1)
    (plus1 n1 n2)
    ...
    (plus1 n39 n40)
    )

  (:goal (and (demand_chilled_goods GV n0)
```

```
            (demand_ambient_goods GV n0)
            (demand_chilled_goods E n0)
            (demand_ambient_goods E n0)
            (demand_chilled_goods BW n0)
            (demand_ambient_goods BW n0)
            (at ADoubleRef depot)
            (at BDouble depot)))
)
```

Note that the object names appearing in the `:objects` section are now typed. The two predicates `demand_ambient_goods` and `demand_chilled_goods` are used to track the remaining (i.e., as yet undelivered) demand for each goods category at each location. The `plus1` predicate expresses when the relation $X = Y + 1$ between pairs of quantity objects is true. In our example we specify that $n1 = n0 + 1, \ldots, n40 = n39 + 1$. The goal to deliver all requested goods is then expressed as the remaining demand being zero and all trucks being at the depot.

Object types in a PDDL domain form a hierarchy. The reserved word `object` denotes the top-level type, which encompasses all objects. In our linehaul problem, there are three types of objects of interest: locations (the depot and the customers), trucks, and quantities. Among trucks, we must further distinguish those that are refrigerated. In the domain above, we have declared `location`, `truck`, and `quantity` to all be subtypes of `object`, and furthermore we have declared `refrigerated_truck` to be a subtype of `truck`.

The role of types specifications is to restrict the instantiation of parameters to objects of the right type. That is, instead of creating all possible ground instances of an action schema by substituting the right number of objects, only those where the substituted objects' types are the same as, or a subtype of, the parameters' types, will be considered. Thus, in our domain definition, the `?t` parameter of the `deliver_ambient` action schema can be filled by an object of type `truck` or type `refrigerated_truck` (since the latter is a subtype), while the `?t` parameter of the `deliver_chilled` action schema can only be filled by an object of type `refrigerated_truck`. In this way, we enforce the constraint that chilled goods are only delivered by (and therefore only transported on) refrigerated trucks.

Type declarations for predicate parameters are essentially redundant, because wherever a predicate occurs, whether in an action schema, the initial state specification or the goal, the type of its arguments is given. Declaring more restrictive types for predicate parameters does not limit instantiation of action schemas (but will likely result in an error when passing the domain to a planner or PDDL validator). It may be viewed a good practice to type predicate parameters as a declaration of the domain designer's intent, but it can equally be regarded as an unnecessary source of errors.

The PDDL syntax for specifying the type of a parameter or object is by placing "`- <typename>`" after it. This can be done for parameters in predicate and action declarations in the domain, and in the `:objects` section of the problem definition. Note that the specification of the supertype of a type name in the `:types` section also uses the same syntax. A parameter or object that does not have a type specification belongs only to the top-level type `object` by default. Several parameters, or objects, can be typed with a single type specification by having multiple parameters to the left of the—symbol, as for example in the two delivery actions where the last four parameters are all specified to be of type `quantity`.

This is also an easy mistake to make: writing a declaration like

```
:parameters (?anything ?a_truck - truck)
```

intending the first parameter to be any type of object and only the second to be of the specified type. However, this declaration makes both parameters have type `truck`. If typed and untyped declarations are mixed in the same list (parameters, objects, or types), the untyped elements must be last in the list. Thus, the declaration

```
:parameters (?a_truck - truck ?anything)
```

does achieve the intended effect. Another way to accomplish the same is to use the reserved type name `object` for the parameter that is meant to be unconstrained:

```
:parameters (?anything - object ?a_truck - truck)
```

PDDL does not require that type names are declared before they are used in the `:types` section. Thus, another way of writing the type declaration section in our example is

```
(:types
  refrigerated_truck - truck
  location truck quantity
)
```

This makes the last three types have the default supertype `object`. As is the case everywhere else in the PDDL language, line breaks have no special significance in the `:types` section.

A Formulation Without Types

Object types are not necessary to distinguish different kinds of objects. We have already seen that we can use *static* predicates—that is, predicates that are not modified by any action—to encode static relations between objects, such as, for example, the `valid_move` and `diff_by_two` predicates over locations to specify the valid Knight's moves, or `plus1` to specify the order of objects representing quantities. The effect of typing, which is to restrict instantiation of action schema parameters, can be achieved using static predicates of one argument. For example, we can write our linehaul domain as follows:

```
(define (domain linehaul_without_types)
  (:requirements :strips)
```

```
(:predicates
  (is_a_truck ?obj)
  (is_a_refrigerated_truck ?obj)
  (is_a_location ?obj)
  (is_a_quantity ?obj)
  ... ; other predicates as above
)


(:action deliver_ambient
  :parameters (?truck ?loc ?d ?d_less_one ?c ?c_less_one)
  :precondition (and (is_a_truck ?truck)
                     (is_a_location ?loc)
                     (is_a_quantity ?d)
                     (is_a_quantity ?d_less_one)
                     (is_a_quantity ?c)
                     (is_a_quantity ?c_less_one)
  ... ; rest of the action schema as above
)


... ; other action schemas rewritten in the same way
)
```

In the problem definition, instead of declaring object types, we specify the instances of the static predicates that are true:

```
(define (problem linehaul-example)
  (:domain linehaul_without_types)

  (:objects
    ADoubleRef
    BDouble
    depot GV E BW
    n0 n1 n2 ... n40
  )

  (:init
    (is_a_truck ADoubleRef)
    (is_a_refrigerated_truck ADoubleRef)
    (is_a_truck BDouble)
    (is_a_location depot)
    (is_a_location GV)
```

```
    (is_a_location E)
    (is_a_location BW)
    (is_a_quantity n0)
    (is_a_quantity n1)
    ... ; rest of initial state as above
    )

  ; goal is unchanged
  (:goal (and ...))
)
```

Note that since there is no defined supertype—subtype relationship here, we must explicitly state that `ADoubleRef` is both a truck and a refrigerated truck.

Since type information can also expressed with static predicates, what are the pros and cons of using PDDL's typing syntax? The main advantage is that it makes intended parameter and object types more explicit, and thereby can help avoiding modelling errors. For example, it is arguably easier to see that an action parameter is missing a type specification—particularly if we adopt the convention of explicitly typing every parameter, including using `object` where no type restriction is intended—than it is to notice that a typing predicate is missing from the action's precondition. Models written using types are also more compact, particularly if deep type hierarchies are used since the subtyping relationship implicitly declares an object to have all of its supertypes. On the other hand, static predicates can be more flexible in expressing properties of objects. For example, if we want to model an object that belongs to more than one type, or a type hierarchy that is not limited to single inheritance, it is unclear how to do this using PDDL's typing mechanism, or indeed whether it can be done at all.

Introducing Action Costs

The PDDL syntax for specifying action costs is actually a restricted case of the same syntax that is used for general numeric functions. (Number-valued functions that are not static are often called *fluents*.) We will cover all of it in Chapter 4 when we describe how to model numeric planning problems. Here, we will introduce only the subset that is permitted for action costs. The specification of action costs in a domain requires three steps:

1. declaring a special numeric function called `total-cost`, with no arguments;

2. adding to the effect of each action an expression that specifies how the action increases the total cost; and

3. adding a `:metric` specification to the problem definition.

Numeric functions are declared in a separate section of the domain definition called `:functions`. In addition to the `total-cost` function, we can also declare static functions, that can be used in the expression that calculates the cost of an action. The values of static functions must be enumerated in the initial state of the problem.

The linehaul domain with action costs is as follows. (We use the formulation with types, and show only the additions to the domain and problem in full. The complete domain and example problem can be found at editor.planning.domains/pddl-book/linehaul_with _costs.)

```
(define (domain linehaul_with_costs)
  (:requirements :strips :typing :action-costs)

  (:types
    ... ;; as above
  )

  (:predicates
    ... ;; as above
  )

  (:functions
    (distance ?l1 ?l2 - location) ; distance between locations
    (per_km_cost ?t - truck) ; per-kilometer cost of each truck
    (total-cost)
  )

  ;; actions deliver_ambient and deliver_chilled are unchanged

  (:action deliver_ambient
    ...
  )

  (:action deliver_chilled
    ...
  )

  (:action drive
    :parameters (?t - truck ?from ?to - location)
    :precondition (at ?t ?from)
    :effect (and (not (at ?t ?from))
                 (at ?t ?to)
                 (increase (total-cost)
```

```
                          (* (distance ?from ?to) (per_km_cost ?t))))
  )
)
```

Note the following changes.

- We have added the keyword `:action-costs` to the `:requirements` section.

- We have added a `:functions` section, declaring two static functions `distance` and `per_km_cost` and the `total-cost` function.

- We have added an `increase` effect to the driving action.

The general syntax of an increase effect is (`increase` *<function term> <expression>*). Because we are defining a domain with only action costs, not general numeric planning, the function term must be `total-cost`. The expression can be made up of numeric constants (integer or decimal), static functions, and the four standard arithmetic operators (+, −, ∗ and /). Same as the logical operators, numeric expressions are written in prefix notation, i.e., with the operator first followed by both arguments. Addition and multiplication can only take two arguments, and there is no unary minus. Note that some planners may only accept more restricted forms of action cost specification, for example, only constants.

The problem definition is extended as follows:

```
(define (problem (linehaul_example)
  (:domain linehaul_with_costs)

  (:objects
    ... ;; as above
  )

  (:init
    ... ;; initial and static facts as above
    ;; specify distances between locations:
    (= (distance depot GV) 573)
    (= (distance depot E) 896)
    (= (distance depot BW) 876)
    (= (distance GV depot) 573)
    (= (distance GV E) 372)
    (= (distance GV BW) 296)
    (= (distance E depot) 896)
    (= (distance E GV) 372)
    (= (distance E BW) 79)
```

```
    (= (distance BW depot) 876)
    (= (distance BW GV) 296)
    (= (distance BW E) 79)
    ;; specify per-kilometer cost for the trucks:
    (= (per_km_cost ADoubleRef) 3.04)
    (= (per_km_cost BDouble) 2.59)
    ;; initialise total cost to zero:
    (= (total-cost) 0)
  )


  (:goal
    ... ; as above
  )


  (:metric minimize (total-cost))
)
```

The syntax for specifying the initial state value of a function, whether it is fluent or static, is (= *<function term>* *<constant>*). Note that the problem must initialise the (total-cost) function to zero.

The final change is the addition of the :metric section, which specifies that the value of the (total-cost) function is to be minimised.

2.2   PLANS AND PLAN VALIDITY

The solution to a planning problem is a plan. In this section, we will define precisely what a plan is, and the conditions necessary for it to solve a planning problem. By doing this, we also give a precise semantics to the fragment of PDDL that we have presented so far.

Let D be a domain definition and P a problem definition for D. We will use the following notation to refer to the components of the domain and problem:

- types(D) the set of type names mentioned in the :types section of D.
- predicates(D) the set of predicates defined in D.
- actions(D) the set of action schemas defined in D.
- For each action schema $a \in$ actions(D), name(a) is the action's name and param(a) is the sequence $x_1, \ldots, x_k$ of its parameters. For each parameter x_i, type-of(x_i) is the type name that the ith parameter of the action is declared to have. If the parameter is declared without a type, type-of(x_i) = **object**. In the same way name(p) and param(p) denotes the name and parameters, respectively, of each predicate $p \in$ predicates(D).
- objects(P) is the set of object names mentioned in the :objects section of P.

- init(P) is the set of ground facts listed in the :init section of P.

- goal(P) is the formula in the :goal section of P.

2.2.1   SEQUENTIAL PLANS

Informally, a plan is a sequence of actions. The actions that appear in a plan are names of ground instances of the action schemas in the domain. Formally, a plan is defined as follows.

Definition 2.1   Let $a \in$ actions(D) be an action schema, with param(a) = $x_1, \ldots, x_k$. Let $o_1, \ldots, o_k$ be a sequence of object names such that $o_i \in$ objects(P) for $i \in 1, \ldots, k$. Then (name(a) o_1 $\ldots$ o_k) is a *ground action name*.

Definition 2.2   A *sequential plan* is a sequence of ground action names.

The following is a plan for the second version of Knight's Tour defined in Section 2.1.2 (Examples 6 and 7). It is the plan that is illustrated in Figure 1.3 in Chapter 1 (page 4).

```
(move_2row_1col n1 n8 n2 n6)
(move_2row_1col n2 n6 n3 n8)
(move_2col_1row n3 n8 n1 n7)
(move_2row_1col n1 n7 n2 n5)
(move_2row_1col n2 n5 n1 n3)
(move_2row_1col n1 n3 n2 n1)
(move_2col_1row n2 n1 n4 n2)
(move_2col_1row n4 n2 n6 n1)
(move_2col_1row n6 n1 n8 n2)
(move_2row_1col n8 n2 n7 n4)
(move_2row_1col n7 n4 n8 n6)
(move_2row_1col n8 n6 n7 n8)
(move_2col_1row n7 n8 n5 n7)
(move_2col_1row n5 n7 n7 n6)
(move_2row_1col n7 n6 n8 n8)
(move_2col_1row n8 n8 n6 n7)
(move_2col_1row n6 n7 n4 n8)
(move_2col_1row n4 n8 n2 n7)
(move_2row_1col n2 n7 n1 n5)
(move_2row_1col n1 n5 n2 n3)
(move_2row_1col n2 n3 n1 n1)
(move_2col_1row n1 n1 n3 n2)
(move_2col_1row n3 n2 n5 n1)
(move_2col_1row n5 n1 n7 n2)
```

```
(move_2row_1col n7 n2 n8 n4)
(move_2col_1row n8 n4 n6 n3)
(move_2row_1col n6 n3 n7 n1)
(move_2row_1col n7 n1 n8 n3)
(move_2col_1row n8 n3 n6 n2)
(move_2col_1row n6 n2 n8 n1)
(move_2row_1col n8 n1 n7 n3)
(move_2row_1col n7 n3 n8 n5)
(move_2row_1col n8 n5 n7 n7)
(move_2col_1row n7 n7 n5 n8)
(move_2col_1row n5 n8 n3 n7)
(move_2col_1row n3 n7 n1 n6)
(move_2row_1col n1 n6 n2 n8)
(move_2col_1row n2 n8 n4 n7)
(move_2col_1row n4 n7 n6 n8)
(move_2col_1row n6 n8 n8 n7)
(move_2row_1col n8 n7 n7 n5)
(move_2col_1row n7 n5 n5 n6)
(move_2row_1col n5 n6 n6 n4)
(move_2row_1col n6 n4 n5 n2)
(move_2col_1row n5 n2 n3 n1)
(move_2col_1row n3 n1 n1 n2)
(move_2row_1col n1 n2 n2 n4)
(move_2col_1row n2 n4 n4 n3)
(move_2col_1row n4 n3 n2 n2)
(move_2col_1row n2 n2 n4 n1)
(move_2row_1col n4 n1 n5 n3)
(move_2row_1col n5 n3 n6 n5)
(move_2col_1row n6 n5 n4 n4)
(move_2row_1col n4 n4 n3 n6)
(move_2col_1row n3 n6 n5 n5)
(move_2col_1row n5 n5 n3 n4)
(move_2row_1col n3 n4 n4 n6)
(move_2row_1col n4 n6 n5 n4)
(move_2row_1col n5 n4 n6 n6)
(move_2col_1row n6 n6 n4 n5)
(move_2row_1col n4 n5 n3 n3)
(move_2col_1row n3 n3 n1 n4)
(move_2col_1row n1 n4 n3 n5)
```

> **Tool support: Plan validation**
>
> The VAL tool suite (available at `https://github.com/KCL-Planning/VAL`) also in-
> cludes a plan validator, called `validate`. It takes as input the domain and problem,
> and one or more plan files:
>
> `$ validate domain-file.pddl problem-file.pddl plan-file-1 ...`
>
> The plan file can be in several different formats, including one with time-stamped
> actions which is used for validation of plans in temporal and hybrid domains (see
> Chapters 5 and 6). For the classical fragment of PDDL, however, a plan is written
> simply as a list of ground action names, one per line, as shown in the example.
>
> Another plan validator is INVAL (available at `https://github.com/patrikhaslu`
> `m/INVAL`). It is used in the same way, but is restricted to validating non-temporal plans
> only. Since some areas of the specification of PDDL are ambiguous, the two validators
> do not always agree on whether a domain and problem is correct, or a plan is valid.
> Using both helps identify the areas where clarification may be needed.

Definition 2.2 states what a plan is. In the next two subsections, we define under what conditions a plan is *valid*, meaning that it is a solution to the planning problem posed by the domain and problem defintion.

Type Correctness

If the domain does not define any type names, all objects have the default type `object`. In this case, there is no restriction on which objects can be substituted for parameters of an action schema to make a ground action instance. If the domain uses typing, however, the first require-ment for a plan to be valid is that the ground actions that make up the plan are instantiated with objects that are of the correct types for the action's parameters. To state this formally, we first have to define precisely the relation between objects and types. Because types in PDDL can form a hierarchy, there is a subtype—supertype relation over types, as well as a mapping from objects to the types they belong to. The combination of both can be expressed as a so-called "is-a" relation.

Definition 2.3   Let IS-A be a binary relation on $\mathrm{objects}(P) \cup \mathrm{types}(D) \cup \{\mathbf{object}\}$ such that:

(*i*) for all $t \in \mathrm{types}(D)$, t IS-A **object**;

(*ii*) for all $o \in \mathrm{objects}(P)$, o IS-A **object**;

(*iii*) if $\ldots\ t_1\ \ldots\ \texttt{-}\ t_2$ appears in the `:types` section of D, then t_1 IS-A t_2;

(*iv*) if $\ldots\ o\ \ldots\ \texttt{-}\ t$ appears in the `:objects` section of P, then o IS-A t;

   (v) IS-A is transitive; and

   (vi) IS-A is the smallest, w.r.t. inclusion, relation that satisfies conditions *(i)–(v)* above.

   If the domain does not define any types, IS-A will simply relate each object to the top-level type name `object`.

Definition 2.4   Let $a \in \text{actions}(D)$ be an action schema with $\text{param}(a) = x_1, \ldots, x_k$ and $n = (\text{name}(a)\ o_1\ \ldots\ o_k)$ a ground action name. We say that n is *type correct* iff o_i IS-A type-of(x_i) for $i = 1, \ldots, k$. A plan is type correct iff every ground action name in it is type correct.

   The use of object typing also places a number of requirements on the domain and problem definitions to ensure they are type correct. We will discuss these in more detail in Section 2.3, but in brief they are as follows. First, note that it is possible to write a PDDL domain definition so that the resulting IS-A relation is cyclic; this is not allowed. Second, for every type correct instantiation of an action schema the arguments of predicates in its precondition and effects must be of the correct type (and number) for those predicates. Finally, the initial state and goal specified in the problem, which are a list of ground literals and a ground formula, respectively, must also be type correct for the predicate parameters.

States and State Transitions
The actions of a plan cause changes to the state. If the plan is successful, the state reached at the end of it is one that satisfies the goal. In the discrete and deterministic fragment of PDDL, a state is defined by the set of facts that are true in it.

Definition 2.5   Let $p \in \text{predicates}(D)$ be a predicate, with $\text{param}(p) = x_1, \ldots, x_k$. Let $o_1, \ldots, o_k$ be a sequence of object names such that $o_i \in \text{objects}(P)$ and o_i IS-A type-of(x_i) for $i \in 1, \ldots, k$. Then $(\text{name}(p)\ o_1\ \ldots\ o_k)$ is a *ground fact*.

Definition 2.6   Let F be the set of all ground facts. A *state s* is a subset of F.

   We interpret a state $s \subseteq F$ as an assignment of a truth value to every ground fact $f \in F$: f is true if $f \in s$ and f is false if $f \notin s$. From this, s gives a truth value to every formula over ground facts, in the usual way: (`not` φ) is true iff φ is false, (`and` $\varphi_1\ \ldots\ \varphi_n$) is true iff each of $\varphi_1, \ldots, \varphi_n$ is true, and so on. The initial state defined by the problem, $\text{init}(P)$, is a state. If the `:init` section of the problem mentions a ground fact that is not a valid instantiation of a predicate of the domain, the problem is not well-formed.
   Recall that a ground action name pairs the name of an action schema, $\text{name}(a)$, with a sequence of object names $o_1, \ldots, o_k$. Each action schema in the domain must have a unique name, so we can find the corresponding schema definition from its name. From the ground action

name, we can define a substitution mapping $\sigma = \{x_1 \rightarrow o_1, \ldots, x_k \rightarrow o_k\}$, where $\mathrm{param}(a) = x_1, \ldots, x_k$. Applying this mapping to the precondition and effects of the action schema definition gives the precondition and effect of the ground action instance.

Definition 2.7 Let $a \in \mathrm{actions}(D)$ be an action schema with $\mathrm{param}(a) = x_1, \ldots, x_k$, $n = (\mathrm{name}(a)\ o_1\ \ldots\ o_k)$ a (valid) ground action name and $\sigma = \{x_1 \rightarrow o_1, \ldots, x_k \rightarrow o_k\}$ the corresponding substitution mapping.

Let $\mathrm{pre}(a)$ denote the `:precondition` formula of a. The *precondition of the ground action*, $\mathrm{pre}(n)$, is the formula $\sigma(\mathrm{pre}(a))$.

Let $\mathrm{effect}(a)$ denote the `:effect` formula of a. The *add effect of the ground action*, $\mathrm{add}(n)$, is the set of ground facts that appear positive (not negated) in the formula $\sigma(\mathrm{effect}(a))$. The *delete effect of the ground action*, $\mathrm{del}(n)$, is the set of ground facts that appear negative (within scope of `not`) in the formula $\sigma(\mathrm{effect}(a))$.

Recall that the effect formula of an action schema must be a conjunction of literals, or a single literal. This restriction ensures that there is no ambiguity about what effects take place when the action is applied (there is no disjunction), and means that we treat the effect formula as a set of literals. The meaning of literals in the effect formula of an action schema is that applying the action makes the positive literals true and the negative literals false, while all facts not mentioned in the effect formula are unchanged. This meaning is captured by the definition of the add and delete effects of a ground action, respectively.

The sequence of ground action names that is the plan inductively defines a sequence of states, which are the states that result from applying each action to the previous state.

Definition 2.8 Let $n_1, \ldots, n_l$ be a plan. The *induced state sequence* $s_0, \ldots, s_l$ is defined inductively by $s_0 = \mathrm{init}(P)$ and $s_i = (s_{i-1} \setminus \mathrm{del}(n_i)) \cup \mathrm{add}(n_i)$, for $i = 1, \ldots, l$.

Definition 2.9 Let $n_1, \ldots, n_l$ be a plan and $s_0, \ldots, s_l$ its induced state sequence. The plan is *executable* iff s_{i-1} satisfies $\mathrm{pre}(n_i)$, for $i = 1, \ldots, l$. The plan is *valid* iff it is executable and s_l satisfies $\mathrm{goal}(P)$.

Notice that in defining the state that results from applying a ground action n, we first remove the facts made false by the action, then add the facts it makes true. This means that if the effect formula of an action schema, under a particular instantiation, contains the same fact both positively and negatively, the net effect will be to make that fact true. However, the presence of the delete effect can still cause *action interference* when discussing non-sequential forms of plans, which is why it is important to distinguish this case from the case where the fact is only made true. Having the add effect take precedence over the delete effect, rather than the other way around, is just a choice made by the designers of PDDL. One should be aware, however, that not all planners implement this semantics; some may consider an action that both adds and deletes a fact to be invalid. From a problem modeller's perspective, the best course may be to avoid writing actions that do this.

We will not show step-by-step the execution of the entire plan for the Knight's Tour problem, but just the first action. The initial state is $s_0 = \text{init}(P) = \{(\text{at n1 n8}), (\text{visited n1 n8}), (\text{diff\_by\_one n1 n2}), \dots, \}$. The set of true instances of the predicates diff_by_one and diff_by_two will be the same in all states reached throughout executing the plan. As described in Section 2.1.2, these are static predicates, which do not appear in the effect of any action.

The first action in the plan is $n_1 = (\text{move\_2row\_1col n1 n8 n2 n6})$. The precondition of this action is $(\text{and} \ (\text{at n1 n8}) \ (\text{diff\_by\_two n8 n6}) \ (\text{diff\_by\_one n1 n2}) \ (\text{not} \ (\text{visited n2 n6})))$, which is satisfied in s_0. The delete effect is $\{(\text{at n1 n8})\}$, and the add effect is $\{(\text{at n2 n6}), (\text{visited n2 n6})\}$. Thus, the state resulting from applying this action to s_0 is $s_1 = \{(\text{at n2 n6}), (\text{visited n1 n8}), (\text{visited n2 n6}), (\text{diff\_by\_one n1 n2}), \dots, \}$.

2.2.2   NON-SEQUENTIAL PLANS

Above, we defined a plan as a sequence of actions. However, there are also less strictly ordered forms of plans. Lifting restrictions on the order of actions is important for scheduling a plan, since it gives flexibility to place the actions in time. However, we will not consider schedules in this chapter. In this section, we will give a brief overview of partially ordered plans. For more details, refer to the articles by Bäckström [1998] and Muise et al. [2016].

For simplicity, we will assume in the following that actions' preconditions are restricted to conjunctions of positive literals. Thus, for a ground action a, we may treat $\text{pre}(a)$ as a set of ground facts. We will also not distinguish between a ground action name and the corresponding ground action schema, as each one can be inferred from the other.

Definition 2.10   Two ground actions, a_1 and a_2, are said to be *non-interfering* iff $(\text{pre}(a_1) \cup \text{add}(a_1)) \cap \text{del}(a_2) = \emptyset$ and $(\text{pre}(a_2) \cup \text{add}(a_2)) \cap \text{del}(a_1) = \emptyset$.

That is, neither action deletes a precondition or add effect of the other. The significance of non-interference is that if both a_1 and a_2 are applicable in a state s, and the actions are non-interfering, then both can be applied one after the other, in either order, and the resulting state will be the same. This provides the foundation for relaxing the order of actions.

Non-interference does not imply concurrency. That two actions are non-interfering means they can be done in any order, but not necessarily that they can be done at the same time. (For example, cleaning the kitchen and cleaning the bathroom can be considered non-interfering actions, since neither prevents the other from being done later, or undoes the effect of the other. But to do them concurrently needs two cleaners.) Non-interference also does not imply that the order of actions a_1 and a_2 when they appear in sequence can be swapped. The first action may add a previously false fact that is required by the precondition of the second. Two actions are *commutative*, which is the term for "swappable" in this sense, iff they are non-interfering and $\text{add}(a_1) \cap \text{pre}(a_2) = \text{add}(a_2) \cap \text{pre}(a_1) = \emptyset$.

Definition 2.11   A *partially ordered plan*, $(T, \text{act}, \prec)$ consists of three elements: $T = \{1, \dots, m\}$ is a set of *steps*, act is a function from T to ground action names, and $\prec$ is a strict partial order[1] on T.

The meaning is that T is the set of steps to be taken as part of the plan, $\text{act}(i)$ is the action that should be done at step i, and $\prec$ specifies the constraints on the order in which these steps can be done. If $\prec$ does not constrain the order of two steps i and j, we are free to choose the order in which to do them.

Validity of a partially ordered plan can be defined in two ways. Let's start with the simple one: A *linearisation* of a partially ordered plan $(T, \text{act}, \prec)$ is the action sequence $\text{act}(i_1), \text{act}(i_2), \dots, \text{act}(i_m)$, where $i_1, \dots, i_m$ is a permutation of $T = \{1, \dots, m\}$, such that $i_j \not\prec i_k$ for all $j > k$. In other words, a linearisation is a possible choice for how to place the actions corresponding to the steps of the plan in a sequence that is consistent with the ordering constraints. A partially ordered plan is valid iff every linearisation of it is a valid sequential plan (as defined in the previous section).

One may think that all unordered actions in a partially ordered plan must be non-interfering, but this is actually not the case. The plan can be valid even if $\text{act}(i)$ deletes a fact added by $\text{act}(j)$ and i and j are unordered, as long as no other action in the plan, or the goal, depends on this fact being true. However, for any two unordered steps i and j, we must have that $\text{pre}(\text{act}(i)) \cap \text{del}(\text{act}(j)) = \text{pre}(\text{act}(j)) \cap \text{del}(\text{act}(i)) = \emptyset$, since there exist linearisations with $\text{act}(i)$ immediately before $\text{act}(j)$ and linearisations with $\text{act}(j)$ immediately before $\text{act}(i)$, and if one action deletes a precondition of the other at least one of these linearisations will not be valid, since the second action will not be applicable in the state that results after the first.

The second way to define the validity of a partial-order plan rests on the notion of a *causal link*. A causal link is a triple (i, p, j), where i and j are steps in T and p a ground fact, such that $i \prec j$, $p \in \text{add}(\text{act}(i))$ and $p \in \text{pre}(\text{act}(j))$. It records an intention that the precondition p required by step j is achieved by step i. A step k with $p \in \text{del}(\text{act}(k))$ is a potential *threat* to the causal link (i, p, j). The link is *safe* from the threat of step k iff any one of the following conditions hold: (1) $k \prec i$; (2) $j \prec k$; or (3) there exists a step l with $p \in \text{add}(\text{act}(l))$ such that $k \prec l \prec j$. The link is *unthreatened* iff it is safe from every potential threat. If the link is unthreatened, then any step that deletes p either cannot appear between i and j in any linearisation, or there is another step (l) that re-establishes p and that must appear between the step that deleted it (k) and the step where it is needed (j). Note that condition (1) above is actually a special case of condition (3), with step i acting as the re-establishing step. To (re-)define plan validity, we need to extend the partially ordered plan with two "fake" steps, 0 and $m + 1$. These represent the initial state and the goal, respectively. For all $i \in T$, we have $0 \prec i \prec m + 1$. With slight abuse

---

[1]A strict partial order is a binary relation that is irreflexive, transitive, and anti-symmetric.

of notation, we say that $\text{add}(\text{act}(0)) = s_I$ and $\text{pre}(\text{act}(m+1)) = G$, even though these steps are not associated with any action. That is, step 0 adds all facts that are initially true, and step $m+1$ requires all goals to be true. The extended partial-order plan $(T \cup \{0, m+1\}, \text{act}, \prec)$ is valid iff for every step $i \in T \cup \{0, m+1\}$ and every fact $p \in \text{pre}(\text{act}(i))$ there exists an unthreatened causal link (j, p, i). This condition is equivalent to the previous one, that all linearisations are valid. (The equivalence was shown by Nebel and Bäckström [1994], although they expressed it in quite different terms.)

Partially ordered plans are distinct from *partial-order planning*, which is a particular method of searching for a plan (that is, of turning a planning problem into a search space). Partial-order planning generates partially ordered plans, but it is equally possible to obtain a partially ordered plan by other means. In particular, a valid sequential plan can be converted into a valid partial-order plan. This process is known as *deordering* [Bäckström, 1998].

2.3   NOTES ON PDDL'S SYNTAX: THE STRIPS FRAGMENT

The examples presented so far in this chapter have introduced most of the basic features of the subset of PDDL known as "STRIPS", plus its extension to negated literals in action preconditions. (The use of not in action preconditions or goals is considered by many to not be within the STRIPS fragment of PDDL.) In this section, we will mention some features that we have not yet seen, and point out some restrictions that are not obvious, as well as some areas of potential ambiguity.

Lexical Elements: Valid Names, Case, and Whitespace
Owing to its origin in LISP, the lexical structure of PDDL is quite simple: it consists of parentheses, symbols (keywords, operators, names, and parameter names), and numeric literals. As we have mentioned, valid names in PDDL are defined to consist of alphanumeric characters, hyphens (-), and underscores (_), and must begin with a letter [McDermott et al., 1998]. However, many planners implement different, either more permissive or more restrictive, definitions. In particular, some planners may not accept names that include hyphens. It is reasonable to assume that reserved words (such as define, domain, problem, and, not, and others) cannot be used as names, although it has never been clearly specified.

Both keywords and names are case insensitive, so they can be written using any combination of upper and lower case. Whitespace is ignored, except where it serves to separate symbols. In particular, line breaks have no semantic meaning anywhere in PDDL. There is, however, one point of ambiguity: is space required to separate the hyphen in a type specification from the parameter or the type name? Some planners may accept, for example, ?t-truck, ?t -truck, and ?t - truck as equivalent, all declaring a typed parameter ?t, while other planners would read the first form as a parameter ?t-truck with no type specified, the second as a syntax error (due to the symbol beginning with -), and only the third form in the intended way. Inserting spaces around the type specifier is the safer option.

Scope and Name Uniqueness

The first PDDL specification [McDermott et al., 1998] stated that within the scope of a domain and problem, all type, function, predicate, action, and object names must be unique. This of course means uniqueness within each category—that is, that we cannot have, for example, two different actions with the same name but different parameters—but it also means uniqness across all name categories; for example, we cannot have a type and an object with the same name. The only exception to this rule is that the domain and problem names may also be used as names within the domain or problem, because these names exist at different scopes. However, the same specification also states that "an atomic type name is just a timeless unary predicate, and may be used wherever such a predicate makes sense", thus suggesting that types are also implicitly declared as predicates.

While the name uniqueness rules have not been explicitly contradicted in any later PDDL version, they are also not strictly adhered to in existing planner, or even plan validator, implementations.

Disjunctive and Multiple Type Specifications

PDDL has a syntax for specifying a disjunction of types: if $t_1, \ldots, t_n$ are type names, then (either t_1 ... t_n) also denotes a type, which is the union of $t_1, \ldots, t_n$. The interpretation of disjunctively typed parameters is straighforward: for example, `?transport - (either ship horse)` means that the parameter `?transport` can be instantiated by an object whose type is `ship` or `horse`, or any subtype of either of those. However, using a disjunctive type in an object declaration, or as the supertype when declaring a type, has no reasonable interpretation, since it either makes the definition nondeterministic, in the sense that we don't know what type the object is, or changes the meaning of `either` to something that is, arguably, its opposite.

The converse of disjunctive types is multiple inheritance, meaning types having multiple immediate supertypes in the hierarchy, or objects belonging to multiple types. For example, suppose we wanted to declare that `horse - vehicle` *and* `horse - herbivore`? Whether or not such multiple specifications for the same type or object can appear in type or object declarations is unknown—it is neither permitted nor ruled out by any of the existing documents on PDDL.

Disjunctive as well as multiple type specifications do not have wide-spread support in planner implementations, and are best avoided. As illustrated in Section 2.1.3, one can always use static predicates to capture object properties that are not expressed by their type.

Implicit Declarations

PDDL allows objects, optionally with types, to be declared in the domain definition as well as in the problem definition. This is done in a section marked by the keyword `:constants`, which follows the same form as the `:objects` section in the problem.

However, due to common use in collections of planning benchmark problems that predated the creation of PDDL, some planners admit "implicit" object declaration, meaning that object names can be used directly in action schemas without a prior declaration. In fact, since predicates, functions, and objects are all defined by the way they appear in action schemas and in the initial state and goal sections of the problem, one may take the view that all declarations are unnecessary and permit all symbols except actions to be declared implicitly. The advantage of adhering to PDDL's explicit declarations is that it makes possible additional validation of the domain, such as checking consistency between the declaration of, for example, a predicate and its use in action schemas. This can help detect modelling errors that would otherwise manifest in ways that make them much harder to debug.

2.4   ADVANCED MODELLING EXAMPLES

In this section we will present a number of slightly more complex examples of problems and how they can be modelled in the subset of PDDL that we have introduced so far. While the examples in Section 2.1 served to introduce the language, the purpose of the following examples is to demonstrate its potential, and in particular some common modelling "tricks" that may not be obvious.

The first example, in Section 2.4.1, shows how we can model a complex action as a sequence of simpler steps, and define actions for each of the steps such that they must always occur in a plan as sequence that makes up an instance of the original action. The second example, in Section 2.4.2, shows we can model a complex goal condition by separating the plan into two distinct phases and verifying the goal condition one simple part at a time.

2.4.1   SORTING BY REVERSALS

Suppose we are given a cyclic permutation of the numbers $1, \ldots, n$, and we wish to sort it into the identity permutation, so that 1 is followed by 2, which is followed by 3, etc., up to n which is followed by 1. The operation we can apply to change the permutation is to take a segment, of arbitrary length, and reverse it. It is always possible to sort the permutation using at most $n - 1$ reversals, but what is the smallest number of reversals? Figure 2.3 shows an example of a permutation of size 5 sorted by two reversals.

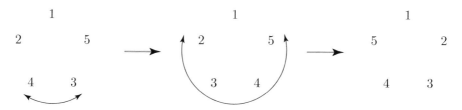

Figure 2.3: Example of sorting a cyclic permutation by reversals.

This problem is related to the problem of comparing genomes [Hannenhalli and Pevzner, 1995]. A genome is an ordered collection of genes. Simple genomes, like those of bacteria or mitochondria, take the form of single cyclic permutation. While insertion, deletion, and substitution of individual nucleotides mutates the genes, changes to their arrangement in the genome at evolutionary scale appear as reversals and transpositions. Because DNA is directional, genomes are usually represented as signed permutations. Here, however, we will only describe a way to model sorting of an unsigned cyclic permutation by reversals. A complete PDDL model of the rearrangement of signed genomes by reversals and transpositions is described by Haslum [2011], and can be found at `editor.planning.domains/pddl-book/ged`.

First, let's consider how to represent states. Given n objects, and we can represent a cyclic permutation with a single predicate (`next ?x ?y`), meaning that `?y` is the next element after `?x`. The direction in which we enumerate the elements of the cycle does not matter; we will assume it is done clockwise. Thus, the leftmost permutation in Figure 2.3 is represented by the facts

```
(next n1 n5) (next n5 n3) (next n3 n4) (next n4 n2) (next n2 n1)
```

For this to represent a cyclic permutation, for every element i there must be exactly one element j such that (`next i j`) is true. This property is another example of an intended *state invariant* of the problem, and we must define the actions of the domain so that they preserve it.

Defining an action that reverses a segment of any fixed length k is simple. For example, the following action reverses a segment of length 2, that is, two adjacent elements:

```
(:action reverse-2
 :parameters (?before_x ?x ?y ?after_y)
 :precondition (and (next ?before_x ?x)
                    (next ?x ?y)
                    (next ?y ?after_y))
 :effect (and (not (next ?before_x ?x))
              (not (next ?x ?y))
              (not (next ?y ?after_y))
              (next ?before_x ?y)
              (next ?y ?x)
              (next ?x ?after_y))
 )
```

Note that as long as there are at least 3 elements, the invariant property that the state is a cyclic permutation guarantees that `?x`, `?y`, and `?before_x`/`?after_y` are distinct (though if there are only three elements, `?before_x` and `?after_y` will be the same). Thus, the action deletes the fact (`next ?x i`), which must be true when the action is applied, and adds a different fact (`next ?x j`), and the same for `?y` and `?before_x`, so it preserves the invariant property.

What makes the general problem more difficult to model is that the size of the segment to be reversed can be arbitrarily large. It is of course bounded by the number of elements in the problem, but if we were to write a different action for each size, we would need a different domain definition for each size of problem. These actions would also have a large number of parameters, which can degrade the efficiency of planners (we discuss this issue further in Section 2.5.6).

Our solution to this problem is to break the action of reversing a segment into a sequence of steps, such that each step affects only a fixed number of elements, and defining an action for each step. We can think of these steps as implementing an "algorithm" for reversing a segment of any length. The algorithm that we will use is as follows.

1. Select two elements *first* and *last*, marking the beginning and end of the segment to be reversed.

2. While *first* is not equal to *last*,

   2(a) Let *before-first* and *after-first* be the elements before and after first, respectively, in the current permutation.

   2(b) Remove element *first* from its current position and insert it after *last*.

   2(c) Change *first* to refer to *after-first*, and repeat step 2.

The steps of the algorithm are illustrated in Figure 2.4 with an example of reversing the segment 5–4–3–2.

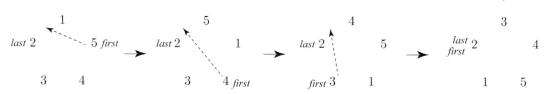

Figure 2.4: Illustration of the step-by-step process for reversing a segment of arbitrary length.

In the PDDL model, each step of the algorithm must be taken by an action, and we have to define the actions such that in any valid plan, they can only appear in the correct order and as a complete instance of a reversal. This means once a reversal has begun, it must be completed before a new reversal is started. To ensure this, we will use a predicate (idle) that is true exactly when there is no ongoing reversal. Two other predicates, (is_first ?x) and (is_last ?x), will indicate that a reversal is in progress and which elements are the current *first* and *last*. The complete domain definition is as follows:

```
(define (domain sorting_by_reversal)
  (:requirements :strips :equality)

  (:predicates
   (next ?x ?y)   ; ?y is after ?x in clockwise order
   (idle)         ; no ongoing reversal operation
   (is_first ?x)  ; the "first" element of the current reversal
   (is_last ?x))  ; the "last" element of the current reversal

  ;; Begin a new reversal of the segment between ?first and ?last:
  (:action begin_reversal
   :parameters (?first ?last)
   :precondition (idle)
   :effect (and (not (idle))
                (is_first ?first)
                (is_last ?last))
  )

  ;; Move one element, and update is_first:
  (:action move_element
   :parameters (?before_first ?first ?after_first ?last ?after_last)
   :precondition (and (is_first ?first)
                      (is_last ?last)
                      (not (= ?first ?last))
                      (next ?before_first ?first)
                      (next ?first ?after_first)
                      (next ?last ?after_last))
   :effect (and (not (next ?before_first ?first))
                (not (next ?first ?after_first))
                (next ?before_first ?after_first)
                (not (next ?last ?after_last))
                (next ?last ?first)
                (next ?first ?after_last)
                (not (is_first ?first)) ; update is_first
                (is_first ?after_first))
  )

  ;; End the reversal when "first" and "last" are the same:
```

```
(:action end_reversal
 :parameters (?element)
 :precondition (and (is_first ?element)
                    (is_last ?element))
 :effect (and (not (is_first ?element))
              (not (is_last ?element))
              (idle))
 )


 )
```

PDDL Example 8: The sorting-by-reversals domain.

Note that we need an inequality to ensure that the `move_element` action can not be applied past the end of the reversed segment.

Because each segment reversal in this model consists of a number of actions, which varies with the length of the segment, the length of the plan does not equal the number of reversals. To actually solve the problem of finding the minimum number of reversals needed needed to sort a given permutation, we need to introduce action costs, such that the cost of beginning a new reversal outweighs the cost of the other actions sufficiently to ensure that the plan with the fewest reversals has the smallest cost. One way to achieve this is to assign a cost of 1 to `begin_reversal` and 0 to the other actions.

2.4.2   DEADLOCK DETECTION

A *deadlock* is a situation in which the execution of a collection of concurrent processes cannot progress because they are all waiting for one another. This kind of situation can arise in concurrent computer systems if they are not designed to avoid such behaviour. If the concurrent processes are modelled as a collection of finite state machines then the question of whether the system can reach a deadlocked state is an example of a model checking problem [Clarke et al., 1993]. As mentioned in Section 1.4.2, planning and model checking are closely related, and here we will show how deadlock detection can also be modelled as a planning problem. The problem that we formulate is to find a plan for the system to reach a deadlocked state. To prove that a system is free of (reachable) deadlocks, we thus need to use a complete planner that is able to prove no plan exists for the problem instance.

To illustrate, we will use the "dining philosophers" problem [see, for example, Hoare, 1985]. This is an artificial example, often used to explain the concept of a deadlock because of its simplicity. *n* philosophers are seated for dinner at a round table, and between each pair of philosophers is one fork. The philosophers' default state is to be thinking, but at any time any one of them can decide to eat. He will then pick up the two forks to either side, eat for a while and then return the forks, making them available for his neighbours to use. Importantly, each philosopher picks up the fork to his right first, and then the fork to his left. A deadlock can occur because once a philosopher has picked up the right fork, he is committed to picking up both forks, and will not return the one he is holding until he has done so. The deadlock occurs if all philosophers pick up their right-hand fork, since they are then all waiting for their left neighbour to return a fork, but none is able to do so. Figure 2.5 shows a philosopher and the adjacent forks as a set of partially synchronised automata. The forks in the example represent resources with mutually exclusive access. In a concurrent computer system, these could be file locks, network access tokens, or some other resource.

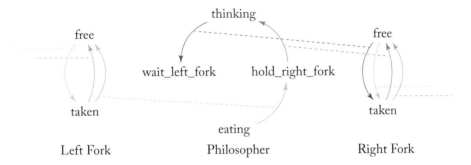

Figure 2.5: Finite automata representing a dining philosopher and the two adjacent forks. Synchronised transitions are linked with dashed line, and coloured the same. Transitions of the two forks are analogously synchronised with the neighbouring philosophers to the left and right.

Modelling the transitions of the automata as action schemas in PDDL is straightforward. The difficulty of this problem is how to model the goal. A deadlock state is one in which no action is applicable. Thus, the goal condition can be expressed as the conjunction over all actions of the negation of their preconditions. This, however, does not fall within the restricted form that goal formulas must have in the STRIPS subset of PDDL.

The trick that we will use to model the problem is similar to that of the previous example, formulating a set of actions that each verifies a simple part of the goal condition. For this, we will need to split the plan into two distinct *phases*: actions in the first phase bring the system to a deadlocked state, while actions in the second phase verify the goal condition. Once the goal-checking phase has begun, no actions that change the state can be taken. This ensures that all parts of the goal are verified in the same state. The separation of the two phases is controlled by a predicate, say (normal), which is set to true in the initial state, and is a precondition of all normal (state-changing) actions. The goal-verifying actions delete this predicate, and after it has been deleted no action can make it true again. This ensures that once the goal-checking phase of the plan has begun, by taking any goal-verifying action, it cannot return to the state-changing phase. The complete domain definitions is as follows:

```
(define (domain philosophers)
  (:requirements :strips :typing)

  (:types philosopher fork)

  (:predicates
   ;; states of a philosopher:
   (thinking ?p - philosopher)
   (wait_left_fork ?p - philosopher)
   (eating ?p - philosopher)
   (hold_right_fork ?p - philosopher)
   ;; states of a fork:
   (free ?f - fork)
   (taken ?f - fork)
   ;; describing the topology:
   (is_right_fork ?p - philosopher ?f - fork)
   (is_left_fork ?p - philosopher ?f - fork)
   ;; predicate to express the goal:
   (deadlocked ?p - philosopher)
   ;; the phase control:
   (normal))

  ;; Actions for the normal (state-changing) phase:

  (:action pick_up_right
   :parameters (?p - philosopher ?f - fork)
   :precondition (and (is_right_fork ?p ?f)
                      (thinking ?p)
                      (free ?f)
                      (normal))
   :effect (and (not (thinking ?p))
                (wait_left_fork ?p)
                (not (free ?f))
                (taken ?f))
  )

  (:action pick_up_left
   :parameters (?p - philosopher ?f - fork)
   :precondition (and (is_left_fork ?p ?f)
```

```
                    (wait_left_fork ?p)
                    (free ?f)
                    (normal))
 :effect (and (not (wait_left_fork ?p))
              (eating ?p)
              (not (free ?f))
              (taken ?f))
 )

(:action put_down_left
 :parameters (?p - philosopher ?f - fork)
 :precondition (and (is_left_fork ?p ?f)
                    (eating ?p)
                    (taken ?f)
                    (normal))
 :effect (and (not (eating ?p))
              (hold_right_fork ?p)
              (not (taken ?f))
              (free ?f))
 )

(:action put_down_right
 :parameters (?p - philosopher ?f - fork)
 :precondition (and (is_right_fork ?p ?f)
                    (hold_right_fork ?p)
                    (taken ?f)
                    (normal))
 :effect (and (not (hold_right_fork ?p))
              (thinking ?p)
              (not (taken ?f))
              (free ?f))
 )

;; Actions for the goal-checking phase:

(:action deadlock_right
 :parameters (?p - philosopher ?f - fork)
 :precondition (and (is_right_fork ?p ?f)
                    (thinking ?p)
```

```
                              (taken ?f)
                              (normal))
   :effect (and (not (normal))
                (deadlocked ?p))
   )

  (:action deadlock_left
   :parameters (?p - philosopher ?f - fork)
   :precondition (and (is_left_fork ?p ?f)
                      (wait_left_fork ?p)
                      (taken ?f)
                      (normal))
   :effect (and (not (normal))
                (deadlocked ?p))
   )

  )
```

The four possible states of a philosopher are represented by predicates (thinking ?p), (wait_left_fork ?p), (eating ?p), and (hold_right_fork ?p). Again, we have a state invariant: for every philosopher, exactly one of these is true in every reachable state.

The two actions for verifying that a philosopher is deadlocked correspond to the thinking and wait_left_fork states, since in the other two states he is holding both or one forks and nothing can prevent him from releasing them. Both of them add a fact (deadlocked ?p), which is used to express the goal that all philosophers are in a deadlocked state. The following is a small example problem with three philosophers:

```
(define (problem three)
  (:domain philosophers)

  (:objects Kant Heidegger Wittgenstein - philosopher
            f1 f2 f3 - fork)

  (:init
   (is_right_fork Kant f1)
   (is_right_fork Heidegger f2)
   (is_right_fork Wittgenstein f3)
   (is_left_fork Kant f3)
   (is_left_fork Heidegger f1)
   (is_left_fork Wittgenstein f2)
   (thinking Kant)
```

```
(thinking Heidegger)
(thinking Wittgenstein)
(free f1)
(free f2)
(free f3)
(normal))

(:goal (and (deadlocked Kant)
            (deadlocked Heidegger)
            (deadlocked Wittgenstein)))
)
```

A possible valid plan for this problem is as follows:

```
(pick_up_right Kant f1)
(pick_up_right Heidegger f2)
(pick_up_right Wittgenstein f3)
(deadlock_left Kant f3)
(deadlock_left Heidegger f1)
(deadlock_left Wittgenstein f2)
```

2.5 EXPRESSIVENESS AND COMPLEXITY

Since PDDL is a language for expressing a certain kind of computational problem, the expressivity of the language is closely related to the computational complexity of solving those problems. In this section, we will review a few classical and well-known complexity results for the class of problems expressible in two subsets of the STRIPS fragment PDDL. The two subsets are without and with parameterised predicates and actions. We refer to the subset of the language not using parameters as *ground STRIPS*. As we will show below, using parameters allows for an exponentially more compact encoding of certain problems. Because the computational complexity of a problem is characterised as a function of the size of the problem instance, the introduction of parameters translates into an exponential increase in the worst-case complexity of the general planning problem.

We assume that the reader is familiar with standard computational complexity notions, such as reducibility and complexity classes. For an introduction, see Garey and Johnson [1979]. The complexity classes we will deal with here are only PSPACE and EXPSPACE.

Let D be a domain definition and P a problem definition for D. We use $|\cdot|$ to denote the size of an object (such as D or P). We distinguish the following computational problems.

- Plan validation: Given D, P, and a plan π, decide whether π is valid for D and P.

- Plan existence: Given D and P, decide whether there exists a plan that is valid for D and P.

- Plan generation: Given D and P, output a plan that is valid for D and P, or determine that no such plan exists.

- Optimal plan cost (generation): Given D and P, output the minimum cost of any plan (resp. a minimum-cost plan) for D and P, or determine that no such plan exists.

These problems are clearly related. For example, if we can solve (optimal) plan generation, then we can also answer plan existence at no more than the same computational cost.

2.5.1   PLAN VALIDATION

The first observation is that plan validation is polynomial in $|D|$, $|P|$, and $|\pi|$. To validate the plan, we need to verify that each ground action name in π is a valid instance of an action schema in D, construct each state in the induced state sequence, and test the applicability of each action in its respective state. To do this, we only need to keep track of the "current" state, meaning the one reached after executing the first k actions in the plan. Once the whole plan has been executed, we can test whether the final state satisfies the goal formula.

Finding the action schema in D that a ground action name corresponds to and checking that it has the right number of arguments for the parameters of the schema can be done in time linear in $|D|$. To check that the instantiation is type-correct, we need to compute the IS-A relation, which requires at worst time cubic in $|\,\mathrm{types}(D)|$ (to compute the transitive closure) and linear in $|\,\mathrm{objects}(P)|$. Given a state s, evaluting a ground formula φ in s can be done in time linear in $|\varphi| \times |s|$ (using time $|s|$ to decide the truth of each ground fact in φ). It remains only to show that the size of the current state will remain bounded. The initial state is part of the problem definition, and therefore bounded by $|P|$. The number of facts added by each ground action in the plan must be less than $|D|$. Thus, the size of each state in the induced state sequence is bounded by $|D| \times |\pi|$.

2.5.2   BOUNDS ON PLAN LENGTH: GROUND STRIPS

In a domain and problem without parameters, each predicate is a single ground fact. Thus, the size of the set of possible ground facts is at most $m = |\,\mathrm{predicates}(D)|$, which means the number of distinct possible states is at most 2^m.

If there is a valid plan for D and P, then there is also a valid plan of length at most $2^m - 1$. Suppose that $\pi = n_1, \ldots, n_l$ is a valid plan such that $l > 2^m - 1$. The length of the induced state sequence is then greater than 2^m, which means it must contain at least one repeated state. Suppose that states i and j are the same: then $\pi' = n_1, \ldots, n_i, n_{j+1}, \ldots, n_l$ is also a valid plan for D and P. Because the state s_i reached by applying $n_1, \ldots, n_i$ is the same as the state s_j reached by applying $n_1, \ldots, n_j$, action n_{j+1} is applicable also in s_i, and applying it to s_i results in a state that is again the same as the state that results from applying it to s_j. Thus, the whole suffix $n_{j+1}, \ldots, n_l$ can be applied from state s_i, and results in the same final state as the origial plan π. Since π is valid, the goal formula holds in this state. The length of π' is strictly less than that of π. Thus, π can be replaced with successively shorter valid plans, until it is not longer than $2^m - 1$.

Next, we will show that $2^m - 1$ is also a lower bound on plan length, that is, that there exist ground STRIPS domains and problems which require exponential-length plans. The simplest example of such a domain and problem is one that models a binary counter. The domain has m predicates, (bit_0), (bit_1), ..., (bit_($m - 1$)), and m actions, (set_0), ..., (set_($m - 1$)). Action (set_i) has the precondition (and (bit_0) ... (bit_($i - 1$))) and the effect (and (bit_i) (not (bit_0)) ... (not (bit_($i - 1$)))). Action (set_0) has no precondition, and only the effect (bit_0). In other words, the action that sets the ith bit to 1 (true) requires all lower-indexed bits to be set, and resets them to 0 (false). The initial state is empty (all facts are false) and the goal formula requires all m predicates to be true. We can show by induction that to reach a state in which (bit_0),...,(bit_i) are true requires $2^{i+1} - 1$ actions. As the base case, to make (bit_0) true requires only one action, and $2^1 - 1 = 1$. To make (bit_0),...,(bit_i) all true requires first making (bit_0),...,(bit_($i - 1$) true, which, by inductive assumption, requires $2^i - 1$ actions. We can then apply action (set_i) which adds (bit_i), but because this action deletes (bit_0),...,(bit_($i - 1$), another $2^i - 1$ actions are needed to make those facts true again. The total plan length is $(2 \times 2^i - 1) + 1 = 2^{i+1} - 1$.

2.5.3 COMPUTATIONAL COMPLEXITY OF GROUND STRIPS PLANNING

The plan length bounds above have as an immediate consequence that the (optimal) plan generation problem requires, in the worst case, exponential time, simply because the plan that is generated may be exponentially long.

The plan validation procedure together with the observation that in ground STRIPS the size of the set of possible ground facts is linearly bounded by the domain size, given that the plan existence problem can be solved in nondeterministic polynomial space, by a guess-and-check algorithm. The algorithm maintains, using polynomial space, only the current state, and nondeterministically selects actions to apply until the state satisfies the goal formula. Since NPSPACE is equal to PSPACE [Savitch, 1970], this means the plan existence problem for ground STRIPS is in PSPACE.

This problem is also PSPACE-hard [Bylander, 1991]. A reduction can be made directly from the acceptance problem for a polynomially space-bounded Turing machine.

2.5.4 COMPUTATIONAL COMPLEXITY OF PARAMETERISED STRIPS PLANNING

PDDL with parameters allows certain problems to be expressed exponentially more compactly than in the corresponding ground language. If there are $n = |\text{objects}(P)|$ objects, a predicate with k parameters can potentially be instantiated into n^k ground facts. Hence, the number of possible ground facts is already exponential in $|D| + |P|$, and the number of possible states is doubly exponential in the size of the domain and problem. By the argument above, if there exists any plan for a given domain and problem there also exists a plan of length at most 2^{n^k}, where k is the maximum arity of predicates in the domain, and the plan existence question can be answered by a nondeterministic algorithm using single-exponential space.

To show a matching lower bound on plan length, we will construct a model of a binary counter with 2^k bits, which can count up to 2^{2^k}, and construct a problem whose plan requires incrementing the counter at least $2^{2^k}/2$ times. We will use a predicate with $k+1$ parameters, (bit ?n$_0$?n$_1$ ... ?n$_{k-1}$?v), and just two objects, zero and one. The idea is that the first k parameters encode a number $i \in 0, \ldots, 2^k - 1$ in binary, and the last parameter is the value of the ith bit of the counter.

The action of incrementing this counter is complex, and we will need to encode it using a sequence of simpler steps. The procedure for incrementing the counter is as follows: let i denote an index, and $i_0, \ldots, i_{k-1}$ denote the binary representation of i. First, set $i = 0$. As long as the ith bit of the counter is 1 (i.e., (bit i_0 ... i_{k-1} one) is true), reset that bit to 0 and increment i by one; when the ith bit is 0, set it to 1 and reset the index to $i = 0$. To encode this procedure, we will need predicates storing the binary representation of the index i. Because the index needs only k bits, we can do this with k predicates, as in the example of the ground binary counter above. However, because we will need to use the values of the index bits as arguments to the bit predicate we adopt a slightly different representation of this state variable: we use a unary predicate (index_j ?v) for $j \in 0, \ldots, k - 1$, and define the domain such that (index_j zero) is true in any state where $i_j = 0$ and (index_j one) is true in any state where $i_j = 1$. The actions that iterate through and reset counter bits are defined as follows:

```
(:action clear_counter_bit_and_inc_index_j
 :parameters (?n_(j + 1) ... ?n_(k − 1))
 :precondition (and (index_0 one)

                    . . .

                    (index_(j − 1) one)
                    (index_j zero)
                    (index_(j + 1) ?n_(j + 1))

                    . . .

                    (index_(k − 1) ?n_(k − 1))
                    (bit one ... one zero ?n_(j + 1) ... ?n_(k − 1) one))
 :effect (and ((not (bit one ... one zero ?n_(j + 1) ... ?n_(k − 1) one))
             (bit one ... one zero ?n_(j + 1) ... ?n_(k − 1) zero)
             (not (index_0 one))
             (index_0 zero)

             . . .

             (not (index_(j − 1) one))
             (index_(j − 1) zero)
```

```
                (not (index_j zero))
                (index_j one))
              )
      )
```

There is one such action for each $j \in 0, \ldots, k-1$. The action that sets a counter bit is defined as follows:

```
(:action set_counter_bit_and_reset_index
 :parameters (?n_0 ... ?n_(k - 1))
 :precondition (and (index_0 ?n_0)
                    . . .
                    (index_(k - 1) ?n_(k - 1))
                    (bit ?n_0 ... ?n_(k - 1) zero))
 :effect (and (not (bit ?n_0 ... ?n_(k - 1) zero))
              (bit ?n_0 ... ?n_(k - 1) one)
              (not (index_0 ?n_0))
              (index_0 zero)
              . . .
              (not (index_(k - 1) ?n_(k - 1)))
              (index_(k - 1) zero))
      )
```

This action also resets the index to zero, so that the next iteration of the counter increment procedure can begin. Note that in this action's definition, we exploit that PDDL interprets an action effect that both adds and deletes a fact as making the fact true. This allows us to delete the currently true fact (`index_j ?n_j`) and add (`index_j zero`) without needing to create different cases for what the current value of each index bit is.

There is one more complication: to initialise the counter to zero, we need to set (`bit` i_0 ... i_{k-1} `zero`) to true for every index $i \in 0, \ldots, 2^k - 1$. Writing these facts explicitly in the `:init` section would make the size of the problem definition exponential in k, and therefore invalidate the claim that the plan is doubly exponential in the size of the domain and problem. To overcome this issue, we can define a set of actions that iterate through the values of the index and add facts (`bit` i_0 ... i_{k-1} `zero`) in the same way as we defined the actions above to clear the counter bits. To make sure that these actions are not interleaved with those that increment the counter, we again use a separation of the plan into two phases, like we did in the model of the deadlock detection problem in Section 2.4.2. Here, the two phases are the initialisation followed by a "normal" plan phase. We use two predicates, (`initialising`) and (`counting`), to represent the current phase, and add (`initialising`) to the precondition of the actions that initialise the counter bits and, (`counting`) to the precondition of the actions that implement the counter increment procedure, making sure that each set of actions is applicable only in the correct phase. We also define an action end_init_phase that is applicable when the initialisation phase is complete, and which switches to the couting phase by deleting (initialising) and adding (counting). In this way, the :init section of the problem must contain only the facts (initialising) and (index_j zero) for $j \in 0, \ldots, k-1$. The :goal of the problem is (bit one ... one one), that is, for the highest bit of the counter to be set. To achieve this goal requires applying the action set_counter_bit_and_reset_index $2^{2^k}/2$ times.

A complete domain and problem definition for $k=4$ can be found at editor.planning.domains/pddl-book/counter4. Note that although the problem only requires the counter to go up to 32,768, the step-wise increment actions and the initialisation phase mean that the shortest plan has 65,551 actions. This, however, does not mean that this plan is difficult to find, since in every reachable state of the problem there is only one applicable action. What the example demonstrates is just that it is possible for plans to be very long.

Using the same principle as in this example, a reduction like that which proves PSPACE-hardness of the ground STRIPS problem but exponentially more compact can be made through the use of parameters. In this way, it was shown that the plan existence question for the general STRIPS language is EXPSPACE-hard [Erol et al., 1991]. We will not repeat the details of the reduction here.

2.5.5 REVERSAL AND COMPLEMENT

The STRIPS fragment of PDDL has two intriguing closure properties: it is closed under *reversal*, meaning that for every STRIPS planning problem P there exists another problem P^R such that plans for P^R are exactly the plans for P in reverse, and it is closed under *complement*, meaning that for every STRIPS planning problem P there exists another problem $\overline{P}$ such that a plan exists for $\overline{P}$ if and only if no plan exists for P.

The reversibility of STRIPS planning was first demonstrated by Massey [1999]; a simplified construction of the reversed problem was proposed by Suda [2016]. We will not recount the details here.

The complementarity of plan existence for the STRIPS formalism is a consequence of the problem's computational complexity. Consider ground STRIPS: the plan existence problem is in PSPACE. PSPACE is closed under complement, which means the plan non-existence problem is in the same class, and thus there exists a polynomially space-bounded Turing machine that recognises precisely the set of unsolvable ground STRIPS planning problems. Since ground STRIPS planning is also PSPACE-hard, the acceptance problem of this Turing machine can be reduced to a ground STRIPS planning problem, via Bylander's [1991] reduction. The same argument can be applied to parameterised STRIPS which is EXPSPACE-complete, since EXPSPACE is also closed under complement.

2.5.6   PRACTICAL CONSIDERATIONS

The complexity results described above concern the worst case, in the sense of the hardest problems that can be expressed in the grounded and parameterised STRIPS subset of PDDL. When modelling a particular problem with the aim of solving it with a planner, however, the more relevant question is how to formulate the problem in a way that is most favourable to the planner, or at least does not unnecessarily disadvantage it. Although planners are intended to be domain-independent problem solvers, it must be recognised that the formulation of the domain can have a significant impact on the efficiency of planners. In other words, the question is how to write *efficient* PDDL.

There are no simple rules for how the domain formulation affects the efficiency of a planner, apart from the rule that *it depends on the planner*. Different PDDL planners have been implemented using a wide range of problem-solving techniques, which are affected in different ways. Yet, there are a few principles that are fairly generally applicable, which we will briefly discuss in the rest of this section.

Grounding

Although all PDDL planners accept parameterised domains as input, the majority internally convert this into a parameter-free, or grounded, representation, by instantiating actions and predicates with all valid combinations of arguments. This is done because the grounded representation lends itself to simpler and more efficient implementation of many solving techniques. However, since the number of ground instances of an action can be exponential in the number of parameters it has, this is a potentially costly operation. (Younes and Simmons [2002] compare the advantage and disadvantage of grounding in the context of a particular planning algorithm.)

As a first illustration, let us consider the different models of the Knight's Tour from Section 2.1.2. As we noted, there are 336 valid knights moves on a regular 8×8 chess board, so if we write a ground PDDL model it will have 336 actions. Our first parameterised domain formulation (Example 4) has a single action schema with two parameters. The problem has 64 objects, one for each square on the board, so there are $64^2 = 4096$ possible ground instances of this action schema. However, recall the static predicate `valid_move`, which is true only of those pairs of squares, listed in the initial state of the problem definition, which are the start and end of a valid Knight's move, and which is a precondition of the `move` action schema. Thus, 3760 of the `move` action instances have a precondition that is statically false. It is not difficult to filter out from the ground instances those with statically false preconditions, and one can safely assume that any good planner implementation does this in an efficient manner. Thus, we can reasonably say that this domain formulation is no more difficult than the ground version. Our second domain formulation (shown in Example 6) has two action schemas with four parameters, but only eight objects, leading to potentially $2 \times 8^4 = 8192$ instances, but again the static predicates in the preconditions of those action schemas limit the ground actions whose preconditions are not statically false to the same number.

As a second example, consider the problem of sorting by reversals presented in Section 2.4.1. The action schema for reversing a segment of fixed length two has four parameters, leading to n^4 potential instances on a permutation of size n. Unlike the previous example, this action schema has no static preconditions. The preconditions that force the four arguments to be a contiguous subsequence of the permutation mean that no more than n ground actions are applicable in any given state, but for every one of the n^4 possible instantiations there is some reachable state in which that ground action is applicable. Thus, a planner that grounds the problem is forced to generate all of them. If we write actions in the same way for reversing each fixed segment length up to $n-1$, we end up with more than n^{n-1} ground action instances, which will quickly become infeasible. The step-by-step domain formulation, shown in Example 8, in contrast, has a set of action schemas, with a fixed number of parameters, for any problem size, and thus no more than $n^2 + n^5 + n$ ground instances. (For large values of n, even this may be too much for a planner to handle. It can be reduced by further splitting up the `move_element` action schema into two steps that remove and insert the element, respectively.)

Symmetries in the Space of Plans

It is often the case that there are many different valid plans, even many different optimal plans, for a given planning problem. This can be the result of reordering of independent actions in the plan, of substituting different but equivalent objects, or arise simply because the problem can be solved in different ways. To illustrate, consider the logistics domain model introduced in Section 2.4, and suppose we have a problem instance with m identical trucks, and a valid plan in which only $k < m$ trucks are used; each truck takes l actions. There are $\binom{m}{k}$ different choices of which k trucks to use; for each choice of k trucks, there are $k!$ different assignments of the chosen trucks to the planned routes; and finally, for each assignment of trucks to routes, there are $\frac{(kl)!}{(l!)^k}$ different interleavings of their actions into a sequential plan. All these plans are equivalent, having the same length and cost.

In some cases, it is possible to rewrite the domain model to eliminate some of the symmetry. In the example of the logistics domain, we can define an ordering of the trucks and enforce that `drive` actions in the sequential plan occur in that order, i.e., that all `drive` actions involving the first truck must take place before any `drive` action involving the second truck, and so on. Since this serves only to prune different interleavings of independent actions, the ordering of the trucks can be arbitrary. We can also use the ordering to rewrite the domain model so that the possible choice of different subsets of identical trucks that are used in the plan and of the permutation of identical trucks between routes are eliminated.

However, unlike the potential combinatorial explosion due to grounding, which affects the majority of planners, the multiplicity of possible plans presents a problem only for planners based on certain problem solving paradigms, particularly cost-optimising planners that use explicit state-space search [e.g., Helmert and Röger, 2008]. Planners based on, for example, SAT encoding [e.g., Rintanen, 2010] or symbolic state-space search [e.g., Edelkamp et al., 2015], on the other hand may be able to exploit symmetries to compact the encoding of plans or states, so that a domain formulation that breaks the symmetries could potentially decrease their efficiency.

CHAPTER 3

More Expressive Classical Planning

Thus far, we have only considered a simple form of PDDL modelling. While we can represent complex scenarios and behaviour, as evidenced by Section 2.4, the language has been extended in key ways which we discuss here in this chapter. The modifications empower the modeller to succinctly represent complex but common phenomenon in the planning setting, but do not necessarily change the theoretical expressiveness of the language. We discuss these details on expressiveness further in Section 3.5. Sections 3.1–3.2 detail additions to the language that allow for more compact and logical representations of action preconditions and effects (including the ramifications some actions may have), and Sections 3.3–3.4 cover temporal specifications of both soft and hard constraints.

3.1 CONDITIONAL AND QUANTIFIED CONDITIONS AND EFFECTS

Arguably the most pervasive extension to the core PDDL language is the Action Description Language (ADL) [Pednault, 1989]. The core features this introduces include conditional effects (updating the state of the world only in certain situations), universally quantified effects (allowing modellers to refer to a set of objects collectively), disjunctive conditions (allowing a generalised view of state conditions), and both universally and existentially quantified preconditions (allowing for complex action pre-requisites). We will demonstrate each of these in turn through the common running example of an elevator controller.

3.1.1 BASIC ELEVATOR MODEL

In the basic elevator example, we model one or more lifts that must transport people between floors. We start with a basic encoding that uses the features already introduced in Chapter 2, and then progressively add ADL extensions to demonstrate the modelling possibilities. Consider the following domain:

```
(define (domain elevators)
    (:requirements :typing)
    (:types elevator passenger num - object)
```

```
(:predicates
        (passenger-at ?person - passenger ?floor - num)
        (boarded ?person - passenger ?lift - elevator)
        (lift-at ?lift - elevator ?floor - num)
        (next ?n1 - num ?n2 - num)
)

(:action move-up
    :parameters (?lift - elevator ?cur ?nxt - num)
    :precondition (and (lift-at ?lift ?cur) (next ?cur ?nxt))
    :effect (and (not (lift-at ?lift ?cur)) (lift-at ?lift ?nxt))
)

(:action move-down
    :parameters (?lift - elevator ?cur ?nxt - num)
    :precondition (and (lift-at ?lift ?cur) (next ?nxt ?cur))
    :effect (and (not (lift-at ?lift ?cur)) (lift-at ?lift ?nxt))
)

(:action board
    :parameters (?person - passenger ?floor - num ?lift - elevator)
    :precondition (and (lift-at ?lift ?floor)
                       (passenger-at ?person ?floor))
    :effect (and (not (passenger-at ?person ?floor))
                 (boarded ?person ?lift))
)

(:action leave
    :parameters (?person - passenger ?floor - num ?lift - elevator)
    :precondition (and (lift-at ?lift ?floor)
                       (boarded ?person ?lift))
    :effect (and (passenger-at ?person ?floor)
                 (not (boarded ?person ?lift)))
)
)
```

PDDL Example 9: Domain definition for a simple elevator problem.

In this simple domain model, we have types that correspond to elevators, passengers, and floors, and predicates that indicate if the passenger is at a particular floor (`passenger-at`), the passenger has boarded a lift (`boarded`), the location of a lift (`lift-at`), and finally one to represent the order of floors (`next`).

Four actions make up the domain: `move-up`, `move-down`, `board`, and `leave`. While the first two move the elevator in expected ways, the second two moves a passenger to or from the elevator (respectively).

Finally, the following example shows the initial configuration for the problem, with five floors, three passengers (on floors 2, 2, and 4), and two lifts (on floors 1 and 5). The goal is to bring all of the passengers to floor 1.

```
(define (problem elevators-problem)
    (:domain elevators)

    (:objects
        n1 n2 n3 n4 n5 - num
        p1 p2 p3 - passenger
        e1 e2 - elevator
    )

    (:init
        (next n1 n2) (next n2 n3) (next n3 n4) (next n4 n5)
        (lift-at e1 n1) (lift-at e2 n5)
        (passenger-at p1 n2) (passenger-at p2 n2) (passenger-at p3 n4)
    )

    (:goal (and
        (passenger-at p1 n1)
        (passenger-at p2 n1)
        (passenger-at p3 n1)
    ))
)
```

PDDL Example 10: Problem definition for a simple elevator problem.

Tip: Constants in the Domain PDDL

In many PDDL benchmarks, it is common to see objects that would otherwise be specified in a problem file as constants in the domain file. While this goes against the philosophy of separating the domain and problem aspects, it is sometimes required in order to avoid using more complex PDDL expression in the domain file. When this occurs, you will typically see a unique domain PDDL for each and every problem (instead of one domain for all of the problems).

A possible plan for the example elevator problem is as follows:

```
(move-up e1 n1 n2)
(board p1 n2 e1)
(move-down e1 n2 n1)
(leave p1 n1 e1)
(move-up e1 n1 n2)
(board p2 n2 e1)
(move-down e1 n2 n1)
(leave p2 n1 e1)
(move-up e1 n1 n2)
(move-up e1 n2 n3)
(move-up e1 n3 n4)
(board p3 n4 e1)
(move-down e1 n4 n3)
(move-down e1 n3 n2)
(move-down e1 n2 n1)
(leave p3 n1 e1)
```

3.1.2 CONDITIONAL EFFECTS

The first extension we will introduce is *conditional effects*. This allows us to model an effect on the world *only under certain conditions*. The format of a conditional effect is

```
(when <head> <body>)
```

The interpretation of this form is "when *<head>* holds in the current state, *<body>* should hold in the state reached by applying this action". For our running example, we will first introduce a set of constants to refer to the passengers as part of the domain:

```
(:constants p1 p2 p3 - passenger)
```

Now that we have the ability to refer to the individual passengers in the domain file itself, we can replace the board and leave actions with load and unload: while the former pair of actions moved individuals, the latter pair can load and unload the entire elevator. The two new actions are as follows:

```
(:action load
    :parameters (?floor - num ?lift - elevator)
    :precondition (and (lift-at ?lift ?floor))
    :effect (and (when (passenger-at p1 ?floor)
                    (and (not (passenger-at p1 ?floor))
                        (boarded p1 ?lift)))
                (when (passenger-at p2 ?floor)
                    (and (not (passenger-at p2 ?floor))
                        (boarded p2 ?lift)))
                (when (passenger-at p3 ?floor)
                    (and (not (passenger-at p3 ?floor))
                        (boarded p3 ?lift))))
)

(:action unload
    :parameters (?floor - num ?lift - elevator)
    :precondition (and (lift-at ?lift ?floor))
    :effect (and (when (boarded p1 ?lift)
                    (and (passenger-at p1 ?floor)
                        (not (boarded p1 ?lift))))
                (when (boarded p2 ?lift)
                    (and (passenger-at p2 ?floor)
                        (not (boarded p2 ?lift))))
                (when (boarded p3 ?lift)
                    (and (passenger-at p3 ?floor)
                        (not (boarded p3 ?lift)))))
)
```

Note that while this will force *every* passenger to either embark or disembark the lift (thus creating erroneous plans in some situations; an issue we will rectify in subsequent examples), it serves as an example of how multiple actions can be combined into a single one that relies on the current state of the world for choosing the effects that result. An example plan for the newly formatted domain is as follows:

```
(move-up e1 n1 n2)
(load n2 e1)
```

```
(move-up e1 n2 n3)
(move-up e1 n3 n4)
(load n4 e1)
(move-down e1 n4 n3)
(move-down e1 n3 n2)
(move-down e1 n2 n1)
(unload n1 e1)
```

Notice how the plan first loads every passenger into the lift e1 prior to unloading them in a single final step on the first floor.

3.1.3 UNIVERSALLY QUANTIFIED EFFECTS

While we have a more concise model with the conditional effects in the previous example, it comes at the cost of explicitly writing the passengers as constants in the domain file. To remedy this, and to drastically simplify the domain model itself, we can use *universal quantification* in the effects of the load and unload actions. A universally quantified effect has the following form:

```
(forall <parameters> <effect>)
```

The *<parameters>* section is precisely the same as an action's :parameters clause: a list of (possibly typed) variables. The *<effect>* is the same as any action effect we have already seen. The following example shows an updated version of the load and unload actions that takes advantage of this. Note that the domain constants for passengers are no longer required; they are simply represented as objects in the problem definition again:

```
(:action load
    :parameters (?floor - num ?lift - elevator)
    :precondition (and (lift-at ?lift ?floor))
    :effect (and (forall (?person - passenger)
                    (when (passenger-at ?person ?floor)
                        (and (not (passenger-at ?person ?floor))
                            (boarded ?person ?lift)))))
)

(:action unload
    :parameters (?floor - num ?lift - elevator)
    :precondition (and (lift-at ?lift ?floor))
    :effect (and (forall (?person - passenger)
                    (when (boarded ?person ?lift)
                        (and (passenger-at ?person ?floor)
                            (not (boarded ?person ?lift))))))
)
```

Note further that the universally quantified effect has a conditional effect nested within it. In general, conditional effects can only contain a conjunction of atomic effects, while quantified effects may contain both atomic effects and conditional effects. A more general representation that reverses the nesting (i.e., having a quantified effect inside of a conditional effect) can always be re-written without changing the representation size or interpretation.

3.1.4 DISJUNCTIVE AND EXISTENTIALLY QUANTIFIED PRECONDITIONS

Even though modern planners are *generally* good at avoiding redundant or useless actions, the example so far allows for a lift to stop at any floor regardless of whether there are any passengers there or on the lift itself. To disallow this, we will restrict the lift from officially stopping (i.e., using the load or unload actions) on a floor in this situation. As a condition, we would like to say "stop only if there is someone on the lift *or* someone waiting to enter it".

The key part of this statement is the disjunction in both the two parts of the statement, and in the notion of "someone". To handle these, we introduce *disjunctive preconditions* and *existentially quantified preconditions*, respectively. The form of a disjunctive precondition follows the same format as an and clause:

```
(or <condition1> <condition2> ···)
```

The interpretation is that only one of `<condition1>`, `<condition2>`, ··· must hold in order for this disjunctive condition to be satisfied. Similar to the universal quantification in syntax (and the disjunctive condition in spirit), this is the format for an existential condition:

```
(exists <parameters> <condition>)
```

The `<parameters>` clause also follows the syntax of the action's parameter list, and `<condition>` is simply any well-formed formula. The interpretation is that *at least* one of the valid object assignments for the `<parameters>` must cause `<condition>` to hold.

One further improvement that we will make with this iteration is to combine the load and unload actions into a single stop action (ultimately, a lift only opens its doors once while on a floor!). The updated action is as follows:

```
(:action stop
    :parameters (?floor - num ?lift - elevator)
    :precondition (and (lift-at ?lift ?floor)
                       (exists (?person - passenger)
                          (or (passenger-at ?person ?floor)
                              (boarded ?person ?lift))))
    :effect (forall (?person - passenger)
                (and (when (passenger-at ?person ?floor)
                       (and (not (passenger-at ?person ?floor))
```

> **Tip: Remodelling**
>
> When planners cannot handle a specific feature of the PDDL language, there may be ways to remodel the problem to avoid specific features. If an action's precondition is made up of a single or clause with k terms, then a way to remodel the problem is to create k copies of the action and use one term for each as a regular precondition. For existential preconditions, a similar approach can be followed after the grounding has occurred.
>
> While in principle this does allow for a wider range of planners to work with the same problem, it can come at the expense of reduced efficiency given the increased number of actions required.

```
                        (boarded ?person ?lift)))
               (when (boarded ?person ?lift)
                  (and (passenger-at ?person ?floor)
                       (not (boarded ?person ?lift))))))
)
```

Again, notice that we are able to nest the different condition types of `exists`, `or`, etc., as long as it makes sense and follows the above patterns of combining the various operators.

3.1.5   UNIVERSALLY QUANTIFIED PRECONDITIONS AND GOALS

The final PDDL feature we will introduce in this section is universally quantified preconditions or goals. We focus on the former in our running example, as the latter follows the exact same syntax and principle.

The syntax for a universally quantified precondition is the same as a universally quantified effect, with the exception that it appears as a term in an action's precondition (similar nesting to the other conditions introduced thus far is allowed). The interpretation is that an action can only be applied if every combination of objects that match the parameters in a `forall` condition lead to the nested condition being satisfied.

As a grounded example, we will extend the elevator domain one more time so that the final action will be for the elevator to enter a "maintenance-mode", and can only do so when all of the passengers have arrived at their floor and the elevator is empty. The first change to our domain is to introduce two new predicates:

```
(requested ?person - passenger ?floor - num)
(in-maintenance)
```

The first represents the requested floor for a passenger (i.e., their destination), and the second will hold only when the elevator is in maintenance mode. The initial state for the problem is augmented to include the passenger requests:

```
(requested p1 n1)
(requested p2 n1)
(requested p3 n1)
```

We further change the problem's goal for the elevator to be in maintenance mode:

```
(:goal (in-maintenance))
```

Finally, we add the action that ensures the elevator enters maintenance mode only when all of the requests have been processed:

```
(:action enter-maintenance-mode
    :parameters (?lift - elevator)
    :precondition (forall (?person - passenger)
                    (and (not (boarded ?person ?lift))
                        (forall (?floor - num)
                          (imply (requested ?person ?floor)
                                 (passenger-at ?person ?floor)))))
    :effect (in-maintenance)
)
```

The example also shows the final logical operator that can be used for constructing conditions in the ADL variant of PDDL: imply. The condition (imply ϕ_1 ϕ_2) holds when either (1) ϕ_2 holds, or (2) ϕ_1 does not hold. Note that imply should not be confused with when: the former is a logical operator that can only be used in formulas, such as in action preconditions, goals, or, indeed, the condition part of a conditional effect; the latter can only be used in the effects part of an action, where it defines a conditional effect. The two are *not* interchangeable.

In this section we have seen a range of syntactic additions to the base PDDL language. Through the careful combination of quantified conditions and effects, complex behaviour can be modelled with these additional operators far more succinctly.

3.2   AXIOMS

Another important extension of PDDL is state dependent *axioms*, which can be used to derive the truth value of some predicates. So far only actions could affect the truth value of a predicate, but now we are going to distinguish between the *standard predicates* and *derived predicates* whose truth value in a state is inferred through axioms only. Derived predicates can then be used as preconditions, goals, or in the context of action effects only, as actions cannot change their truth value directly.

Axioms can be used to elegantly model complex concepts such as graph reachability, network flow, transitivity closure, and relational properties among objects. As a result, axioms typically simplify action schemas, making the preconditions more readable and avoiding effects that are just logical consequences of the standard predicates. Axioms do not increase the language expressiveness in terms of what can be modeled, but decrease the domain description size and plan length, as compiling away axioms can result in an exponential growth in the number of actions [Thiébaux et al., 2005].

The syntax of an axiom is

```
(:derived <predicate> <condition>)
```

The interpretation of this form is "when `<condition>` holds in the current state, `<predicate>` should hold true in the state". We assume negation as failure to know when the derived predicate is false; that is, when `<predicate>` cannot be derived through any axiom, then `<predicate>` is false.

Axioms can nest rules recursively by using the same `<predicate>` in the `<condition>` formula. For our running example, we will define a rule to derive the static relationship describing when one floor is above/below another. That is, we make the transitive closure of the (next ?n1 - **num** ?n2 - **num**) predicate. We need to first add a predicate to our domain definition:

```
(above ?blw - num ?abv - num)
```

The first parameter of the predicate represents the floor number which is below the floor in the second parameter. The axiom rule to derive the truth of above is added to the domain in the same section where actions are defined:

```
(:derived (above ?blw ?abv - num)
    (or (next ?blw ?abv)
        (exists (?z - num)
            (and (next ?blw ?z)
                (above ?z ?abv)))))
)
```

This axiom shows that conditions can be composed using the recently introduced ADL features. The derived predicate above appears in the head of the rule, and a disjunctive condition with an existential quantifier recursively defines the relationship between pairs of floors. The axiom implies floor ?abv is above if a predicate next explicitly states this relationship, or there is another floor ?z which is right next to ?blw and for which the above derived predicate also holds with respect to ?abv.

As all conditions mentioned in our example are either static predicates or the same derived predicate, then derived predicate above is also static, and defines an invariant which is true in all reachable states from the initial state. This is not the case of all derived predicates as they can be defined in terms of predicates whose truth value is changed by actions. Nevertheless, there are certain restrictions on how axioms can be specified, because we want to make sure the truth value of derived predicates are algorithmically easy to derive, do not incur in significant extra computation, and are uniquely defined. The first rule is that no action can change the truth value of a derived predicate. The second rule states that no derived predicate appears *negated* in the conditions. This rule makes sure that no negative interactions appear from the application of axioms, and a unique fixed point is reachable. Both rules are respected in the axiom we introduced for the elevators domain.

We can now use the derived predicate above to allow elevator movement not only to consecutive floors, but to any floor above or below. We further show that with existential quantifiers we can restrict an elevator to only move to a floor if there is a passenger waiting there, or there is a passenger in the elevator that wants to disembark.

```
(:action move-up
    :parameters (?lift - elevator ?cur ?nxt - num)
    :precondition (and (lift-at ?lift ?cur) (above ?cur ?nxt)
                       (exists (?person - passenger)
                               (or (passenger-at ?person ?nxt)
                                   (and (requested ?person ?nxt)
                                        (boarded ?person ?lift)))))
    :effect (and (not (lift-at ?lift ?cur))
            (lift-at ?lift ?nxt))
)
```

PDDL Example 11: Revised move-up action for the elevators domain using axioms.

Action move-down has the same action schema but the derived predicate parameters are reversed: (above ?nxt ?cur). Action load is modeled as in our previous example, boarding all passengers in current floor using a universal quantifier. Action unload can be extended further to make sure only passengers who requested the floor disembark.

```
(:action unload
        :parameters (?floor - num ?lift - elevator)
        :precondition (and (lift-at ?lift ?floor))
        :effect (and (forall (?person - passenger)
                    (when (and (boarded ?person ?lift)
                          (requested ?person ?floor))
                          (and (passenger-at ?person ?floor)
                               (not (boarded ?person ?lift))
```

```
                                    (not (requested ?person ?floor))))))
)
```

PDDL Example 12: Domain definition for an elevator: extending the unload action.

We finally change the goal to compute a plan where one elevator ends up in the first floor, the other in the top floor, and all passenger requests are served.

```
(:goal (and
        (lift-at e1 n1) (lift-at e2 n5)
        (forall (?person - passenger ?floor - num)
                (not (requested ?person ?floor)))
        ))
```

Axioms represent a powerful tool for domain designers to compactly specify key properties of the domain. The planner support for axioms is limited compared to the vanilla classical planning syntax, however for domains with properties such as actions conditioned on the transitive closure of key predicates, axiom capable planners can be far superior compared to the alternative of rewriting the domain [Miura and Fukunaga, 2017, Thiébaux et al., 2005].

3.3   PREFERENCES AND PLAN QUALITY

So far, we have defined planning as the problem of finding a path that maps the initial state into a goal state that achieves a set of goals. We have seen that the goal is typically a conjunction of predicates in STRIPS, but can be expressed using universal and existential quantifiers using the subset of ADL extensions to the language. In this subsection we are going to introduce the notion of soft goals to define some properties that we would like to achieve, but are not compulsory. In many real-world problems not all goals are achievable and hence instead of not computing a plan at all, we rather try to achieve as many goals as possible. This situation arises when goals are mutually exclusive, or they make the problem computationally harder. For example, the well known Traveling Salesman Problem becomes computationally easy if we make the requirement to visit a city at most once as a preference instead of a requirement. Furthermore, not all preferences need to have the same weight, so we will introduce the syntax to specify the quality of a plan in terms of the preferences or soft goals that have been achieved. If preferences are used, we have to include the requirement flag :preferences in the domain definition.

The syntax of a preference is

```
(preference [<name>] <body>)
```

The *<name>* is optional. The interpretation of preferences depends on their context as they can be defined as action preconditions or as goals, but not as conditions of conditional effects. If they are defined as a precondition, their interpretation is of the form "when the action is applied, if formula *<body>* does not hold, it will incur in a penalty in the overall plan quality". Unless specified, the penalty for each occurrence of an unsatisfied preference is a cost of 1 (we show how to specify non-unit costs below as part of the goal metric). If a preference is specified within the (:goal ...) scope, then the truth value of *<body>* the violation penalty is incurred if the formula does not hold in the goal state. Further, nesting of preferences is not allowed and preferences can only appear inside a conjunction or a universal quantifier. The expression in *<body>* can be formulated using conjunctions, disjunctions, universal, and existential quantifiers. For example, the preference

```
(preference allRequestsServed (forall (?person - passenger ?floor - num)
                                (not (requested ?person ?floor))))
```

is satisfied if all passengers requests are served. If one request is not served it will incur in the same penalty as if no requests are served at all. If we reverse the nesting and write the preference inside the universal quantifier,

```
(forall (?person - passenger ?floor - num)
        (preference requestServed (not (requested ?person ?floor))))
```

then each unserved passenger will incur in a penalty. This statement will create a preference with the name requestServed for every floor a passenger can request. Therefore, note that the interpretation changes depending on the nesting of the universal quantifier. Remember that universal quantifiers are read as conjunction of over all legal instantiations of the quantification.

Heterogeneous penalties can be specified using the syntax

```
(is-violated <name>)
```

where *<name>* is the identifier used in one or more preferences. The interpretation of is-violated is as an integer value representing the number of preferences not satisfied with the identifier *<name>*. It can be understood as a counter. For example, (is-violated allRequestsServed) can be 0 or 1, while (is-violated requestServed) can have any value from 0 to n, where n is the maximum number of possible requests. These counters can be combined through standard arithmetic operators to define the quality of a plan. For example, this is how we would specify a linear weighted combination of the preferences introduced so far:

```
(:metric minimize (+  (* 2 (is-violated requestServed))
                      (* 100 (is-violated AllRequestsServed))))
```

A plan that reaches a goal state satisfying all preferences would have a penalty of 0 according to the expression above. If just one request is not served the penalty is 102, two unserved requests would have a penalty of 104, and so on.

As a full example, let us extend the basic elevator domain from before (Example 9) to express the preference of not filling the elevators with more than five passengers. For doing so we need to create a new predicate to keep track of the number of passengers per lift (current-load ?lift - elevator ?n1 - num). The initial state specifies that the lifts are initially empty. In our 2-elevators example, the predicates (current-load e1 n0) (current-load e2 n0) are included in the (:init ...) section. The board action increases current-load every time a passenger boards a lift, and leave decreases the counter when a passenger leaves. Whenever someone boards a lift, the preconditions capture the preference of having zero to five passengers at most.

```
(:action board
    :parameters (?person - passenger ?floor - num ?lift - elevator
                 ?cur_load - num ?nxt_load - num )
    :precondition (and (lift-at ?lift ?floor)
                       (passenger-at ?person ?floor)
                       (next ?cur_load ?nxt_load)
                       (current-load ?lift ?cur_load)
                       (preference maxLoad (or
                           (current-load ?lift n0)
                           (current-load ?lift n1)
                           (...)
                           (current-load ?lift n4))))
    :effect (and (not (passenger-at ?person ?floor))
                 (not (current-load ?lift ?cur_load))
                 (boarded ?person ?lift)
                 (current-load ?lift ?nxt_load)))

(:action leave
    :parameters (?person - passenger ?floor - num ?lift - elevator
                 ?cur_load - num ?nxt_load - num)
    :precondition (and (lift-at ?lift ?floor)
                       (boarded ?person ?lift)
                       (next ?nxt_load ?cur_load)
                       (current-load ?lift ?cur_load))
    :effect (and (passenger-at ?person ?floor)
                 (current-load ?lift ?nxt_load)
                 (not (boarded ?person ?lift))
                 (not (current-load ?lift ?cur_load))))
```

We can now use the same goal definition as before, in Example 10, and the best plan would not load an elevator with more than five passengers. To realize the effect of the preference we need to add more passengers to the problem as the original definition had only three passengers.

3.4   STATE TRAJECTORY CONSTRAINTS

While we have seen how to encode preferences over goals and action preconditions, also known as soft goals/preconditions (in contrast to hard goals/preconditions that must be satisfied by any valid plan), we now introduce how to specify both hard and soft constraints on entire state trajectories. State trajectory constraints typically express conditions that are not state dependent, but rather depend on the path taken to reach a state [Gerevini et al., 2009]. As such, hard constraints encode properties valid plans must satisfy, while soft constraints encode preferences used to define plan quality.

Constraints are expressed through temporal modal operators, akin to operators in linear temporal logic (LTL) [Pnueli, 1977], over first-order formulae using domain predicates. Constraint formulae cannot express occurrences of actions as they refer to state properties only, which appear as predicates[1]. The five basic modal operators are:

1. (always `<condition>`)

2. (sometime `<condition>`)

3. (at-most-once `<condition>`)

4. (at-end `<condition>`)

5. (within `<num>` `<condition>`)

While the interpretation of LTL formulae is different over finite and infinite plans[2], the interpretation of PDDL modal operators is specified only over finite plans. The first modal operator always enforces a condition to be true over all states in the plan, sometime enforces that the formula is satisfied by one or more states along the plan, very similar to the notion of "at-least-once". The third operator states that the formulae *becomes* true at most once along the plan. That is, a plan can satisfy this condition by never making the formula true, or by making sure that once it becomes true, it can be falsified but not made true again. The fourth modal operator is equivalent to the standard goal conditions we have seen so far, which states that a formulae must hold in the final state of a plan. If a goal formula has no modal operator, we assume that it takes the at-end modality by default. The last modal operator states that the formula is satisfied before `<num>` steps of the plan have been executed from the initial state.

Arbitrary nesting of temporal modalities is not allowed, however, three custom nestings have been introduced to capture commonly used constraints:

6. (sometime-after `<condition1>` `<condition2>`)

---

[1]This restriction is not limiting, as new fluents that represent the occurrence of an action can be introduced, and added to the state only when an action occurs (i.e., as an add effect).

[2]An infinite plan can be created if a loop is added to the plan and all actions in the loop remain always applicable [Patrizi et al., 2011].

7. `(sometime-before` *`<condition1>`* *`<condition2>`*`)`

8. `(always-within` *`<num>`* *`<condition1>`* *`<condition2>`*`)`

The first states that if first condition is true in some state, then future trajectories have to satisfy `(sometime` *`<condition2>`*`)`. The second states that before *`<condition1>`* becomes true `(sometime` *`<condition2>`*`)` must be satisfied. The third states that anytime *`<condition1>`* is satisfied, `(within` *`<num>`* *`<condition2>`*`)` has to hold, taking as reference the state that satisfied the first condition instead of the initial state to count the number of steps.

The final group of state trajectories are those that extend `always` modality over intervals:

9. `(hold-after` *`<num>`* *`<condition>`*`)`

10. `(hold-during` *`<num1>`* *`<num2>`* *`<condition>`*`)`

The first states that `(always` *`<condition>`*`)` must be satisfied from the *`<num>`*th step in the state trajectory onwards. The second states that `(always` *`<condition>`*`)` must hold between the *`<num1>`* and *`<num2>`* steps, including *`<num1>`*.

In order to make a state trajectory constraint soft, we just have to use the same preference syntax introduced previously. The `:constraints` flag has to be added to the domain file and all constraints have to be declared in the `(:constraints ...)` scope, which can be written right after the `(:goal ...)` declaration. For example, the same preference we introduced into the `board` action to prefer plans that do not load more than five passengers per lift, can now be expressed with the following constraint:

```
(:constraints
 (forall (?lift - elevator)
         (preference maxLoad (always (not current-load ?lift n6)))))
```

Standard goals which have to hold true in the final state of a valid plan can be included in the `(:goal ...)` specification or added with the `(at-end ...)` constraint. The goal from the previous example where all passengers had to be moved to floor 1 can be added to the `(:constraints ...)` section as shown below:

```
(:constraints
 (and
  (forall (?lift - elevator)
          (preference maxLoad (always (not current-load ?lift n6))))
  (at-end (passenger-at p1 n1))
  (at-end (passenger-at p1 n2))
  (at-end (passenger-at p1 n3))))
```

We can express rich behavior once we can make use of these modal operators. For instance, we can specify that each passenger can be boarded at most once, to avoid multi-elevator trips:

```
(:constraints
 (forall (?person - passenger ?lift - elevator)
         (at-most-once (boarded ?person ?lift))))
```

We can specify that an elevator should not visit a floor with a boarded passenger more than once, to avoid unnecessary time in the elevator:

```
(:constraints
 (forall (?lift - elevator ?floor - num ?person - passenger)
         (at-most-once (and  (boarded ?person ?lift)
                             (lift-at ?lift ?floor)))))
```

We can even specify that at most one elevator is at the same level at any point in the plan:

```
(:constraints
 (forall (?lift1 ?lift2 - elevator ?floor - num)
         (always (imply (and (lift-at ?lift1 ?floor)
                             (lift-at ?lift2 ?floor))
                     (= ?lift1 ?lift2)))))
```

The range of temporally extended goals allowed for this fragment of PDDL is limited to only the ten we have listed in this section. Nonetheless, combined with the expressive power of the conditions themselves for a single state, the technique represents a powerful tool in the arsenal of a domain modeler to express non-Markovian preferences and constraints.

3.5   EXPRESSIVENESS AND COMPLEXITY

In Section 2.5, we discussed the complexity of the STRIPS subset of PDDL for the computational problems of plan existence, generation, optimal generation, and validation. The first three problems are PSPACE-hard in the ground version and EXPSPACE-hard in parameterized STRIPS, while the problem of validation is linear in the size of the plan. These results hold for the extensions described in this chapter, namely conditional effects, universal and existential quantification, derived predicates, preferences, and finite state trajectory constraints, as none of these features increase the computational complexity of the problems being solved.

However, Nebel [2000] proposed a different way to study the expressiveness of the extensions of PDDL that we have introduced in this chapter, via *compilations* of each extension of the language into STRIPS. A compilation is a problem reformulation that preserves properties such as plan existence, and optimal plan cost. Nebel suggested that if the size of the reformulated problem is at most polynomial in the size of the original problem and the length of valid plans does not increase by more than a constant factor, then the extension does not add expressive power to the language, but rather is "syntactic sugar" allowing modellers to more elegantly capture certain properties of the problem, e.g., allowing negative preconditions and compiling them away with newly introduced fluents to re-achieve the STRIPS property. If every possible compilation of a feature is exponential in problem size, or induces plans lengths in the compilation larger than a constant increase, then the feature adds expressiveness. In the remainder of this section, we will briefly describe compilations of each of the language features introduced in this chapter into the STRIPS subset, to clarify whether they add to the expressiveness of PDDL.

Universal quantifiers can be replaced by a conjunction over the objects of the quantified parameter, and existential quantifiers can be replaced by a disjunction. For example:

```
(forall (?lift - elevator ?floor - num) (lift-at ?lift ?floor))
;Can be compiled into:
(and  (lift-at e1 n1) (lift-at e1 n2) ... (lift-at e1 nx)
      (...)
      (lift-at ex n1) (lift-at ex n2) ... (lift-at ex nx))

(exists (?lift - elevator ?floor - num) (lift-at ?lift ?floor))
;Can be compiled into:
(or  (lift-at e1 n1) (lift-at e1 n2) ... (lift-at e1 nx)
     (...)
     (lift-at ex n1) (lift-at ex n2) ... (lift-at ex nx))
```

These compilations are linear and do not increase the length of a plan. Thus, quantifiers can be considered "syntactic sugar".

A disjunctive action precondition can be compiled away by creating a copy of the action for each disjunct, whose precondition is that disjunct. In any state where the original, disjunctive, precondition holds, at least one of the disjuncts is true and thus at least one of the copies of the action applicable. This, however, only works if the precondition formula is in disjunctive normal form (DNF), meaning it is a disjunction of conjunctions of literals. Any Boolean formula can be rewritten as an equivalent formula in DNF, but doing so may increase the size of the formula exponentially. In fact, preconditions that are general logical formulae cannot be compiled away without either a potential exponential increase in problem size or a super-linear increase in plan length [Nebel, 2000].

Conditional effects can also be compiled away. Two compilation techniques are known: one increases the size of the problem exponentially, but keeps the length of valid plans constant; the other increases the size of the problem only polynomially, but the length of plans more than linearly. It has been shown that there is no compilation that achieves both [Nebel, 2000]. Therefore, conditional effects augment the expressiveness of PDDL in terms of the conciseness of the problem representation. Here, we will only illustrate the first compilation, which is simpler. An action with $C(a)$ conditional effects is replaced by $2^{|C(a)|}$ STRIPS actions, one for every subset of conditional effects, each of whose precondition ensures that it is applicable only in states where exactly that subset of conditional effects would occur. For example, the action

```
(:action unload
    :parameters (?floor - num ?lift - elevator)
    :precondition (and (lift-at ?lift ?floor))
    :effect (and (when (boarded p1 ?lift)
                        (and (passenger-at p1 ?floor)
                             (not (boarded p1 ?lift))))
                 (when (boarded p2 ?lift)
                        (and (passenger-at p2 ?floor)
                             (not (boarded p2 ?lift))))))
```

can be compiled into the following actions:

```
(:action unload-p1
    :parameters (?floor - num ?lift - elevator)
    :precondition (and (lift-at ?lift ?floor)
                       (boarded p1 ?lift)
                       (not (boarded p2 ?lift)))
    :effect (and (passenger-at p1 ?floor)
                 (not (boarded p1 ?lift))))


(:action unload-p2
    :parameters (?floor - num ?lift - elevator)
    :precondition (and (lift-at ?lift ?floor)
                       (not (boarded p1 ?lift))
                       (boarded p2 ?lift))
    :effect (and (passenger-at p2 ?floor)
                 (not (boarded p2 ?lift))))


(:action unload-p1-p2
    :parameters (?floor - num ?lift - elevator)
    :precondition (and (lift-at ?lift ?floor)
                       (boarded p1 ?lift)
                       (boarded p2 ?lift))
    :effect (and (passenger-at p1 ?floor)
                 (passenger-at p2 ?floor)
                 (not (boarded p1 ?lift))
                 (not (boarded p2 ?lift))))
```

Note that the action corresponding to the case when none of the effect conditions is true has been omitted, since it has no effect. In general, this may not be the case since an action may have a mix of conditional and unconditional effects.

The compilation above assumes that the action's effects is a conjunction of "simple" conditional and unconditional effects. However, nested conditional effects can be rewritten as a flat conjunction of conditional effects, and hence are compilable into STRIPS in the same ways. For example:

```
(when <condition1> (and <effect1>
                        (when <condition2> <effect2>)))
```

can be rewritten into the equivalent conjunction

```
(and
  (when <condition1> <effect1>)
  (when (and  <condition1>
              <condition2>) <effect2>))
```

Also note that the increase of model size due to the compilation of conditional effects shown above is only exponential in the number of such effects per action. If, for example, every action has a small number of conditional effects, which does not grow with problem size, this compilation can still be practical. An example of a situation in which it is not is when the domain has quantified conditional effects; although the quantifier can be removed by grounding, this give rises to a number of grounded conditional effects that grow with the number of objects (of the right type) in the problem.

Axioms add expressive power to PDDL as they also reduce the size of the problem. It has been shown that axioms can be compiled away, but that this increases either the problem or plan size by an amount that is exponential in the maximum fixed point depth needed to determine the truth value of a derived predicate [Thiébaux et al., 2005].

Preferences do not increase the expressive power of PDDL, because they can be compiled into STRIPS with action costs efficiently, incurring a linear increase of the size of the problem [Keyder and Geffner, 2009].

For the types of planning problems we have seen so far (where all actions are instantaneous), the state trajectory constraints can be compiled away with only a polynomial increase in the problem size and constant increase in plan length [Gerevini et al., 2009]. Thus, for sequential finite plans, state trajectory constraints do not add a level of expressiveness. When we shift to temporal planning (a fragment of PDDL discussed in Chapter 5), actions may occur concurrently and the compilation increases the plan length linearly, thus causing trajectory constraints to add expressive power to the formalism.

CHAPTER 4

Numeric Planning

In the classical planning problems that we modelled in Chapters 2 and 3, all state variables are instances of predicates which are Boolean-valued. That is, they can only take values true and false. The numeric planning subset of PDDL introduces state variables whose values are (rational) numbers. These are instances of *functions*[1]. The range of a numeric function in PDDL is unbounded, which means there is an infinite set of values it can take. A state is still a set facts, but in addition to those of the form (at Knight n1 n8) we can now have facts of the form (= (row Knight) 8). Although functions can have parameters, these can only be instantiated with objects from the domain or problem. Thus, the number of state variables, and the number of atomic facts (i.e., those without negation) that are true in any state, is still finite—it is only the number of possible (and potentially reachable) states that may now be infinite. Apart from this, the numeric planning fragment of PDDL keeps all the classical planning assumptions: the model is deterministic, the initial state is known, and there are no state changes other than those caused by the plan. Nevertheless, the potentially infinite number of states means that the numeric planning problem is in the worst case only semi-decidable, and that there is no upper bound on the length of a plan in general.

We have already made limited use of functions to define action costs, in Section 2.1.3. In this chapter we present the complete numeric subset of PDDL, which extends the language with expressions and conditions over numeric values (using standard arithmetic and comparison operators, such as $+$, $<$, etc.) and effects on numeric state variables.

As in Chapter 2, we first introduce the new features of PDDL for modelling numeric planning problems through a few examples. In Section 4.2, we then redefine plan validity to account for the numeric aspects of the problem. We present some more advanced modelling with numeric conditions and effects in Section 4.3, and we finish the chapter with a brief discussion of the expressivity and complexity of numeric planning in Section 4.4.

4.1 NUMERIC PLANNING IN PDDL

In Chapter 2 we encountered several models in which we used objects representing a small set of integers. In the following, as we introduce each of the features of the numeric planning fragment of PDDL, we return to the linehaul logistics problem from Section 2.1.3 and show how it can be formulated in numeric PDDL.

---

[1]Mathematically speaking, functions in PDDL are more akin to fluents that take on a numeric value. If static, they may behave like mathematical functions, but should not be conflated.

The requirements keyword for the numeric fragment of PDDL is `:fluents`, which is also the term often used for functions whose value is part of the state, i.e., numeric state variables.

Declaring Functions

Functions are declared in a section of the domain definition named `:functions`. Like predicates, functions can have parameters, which are instantiated with objects (of suitable types) defined in the domain or problem. In the linehaul domain, the quantities we need to model are the remaining, unserved, demand of each customer, per goods type, and the remaining, unused, capacity in each truck. We do this with the following functions:

```
(:functions
  (demand_chilled_goods ?c - location) ; customer's remaining demand for
  (demand_ambient_goods ?c - location) ; chilled/ambient goods
  (free_capacity ?t - truck)           ; unused capacity in truck
  (distance ?l1 ?l2 - location)        ; distance between locations
  (per_km_cost ?t - truck)             ; per-kilometer cost of truck
  (total-cost)
  )
```

Note that we have also kept the function `total-cost`, and the two static functions `distance` and `per_km_cost`, so that we can formulate the objective of minimising total cost. As mentioned in Chapter 2, the syntax for specifying action costs is nothing other than a special case of the general syntax for specifying a metric defined by a numeric expression.

In the numeric fragment of PDDL, all functions are number-valued, so the value type of a function does not need to be specified. PDDL has no mechanism to explicitly restrict the range of a function to particular set of numbers, such as non-negative numbers or integers. When defining numeric planning domains we usually create such restrictions implicitly, however, through the preconditions and effects of actions. This is simply the numeric equivalent of the state invariants we saw in several domains in Chapter 2. In our formulation of the linehaul problem, the values of the PDDL functions (`demand_chilled_goods ?c`), (`demand_ambient_goods ?c`) and (`free_capacity ?t`) can only be decreased, and cannot be decreased below zero since the action preconditions guard against that. Thus, they will be bounded above by their initial values and below by zero.

For compatibility with the extension of PDDL to object-valued functions [Helmert et al., 2008], which we do not describe in this book, it is also possible to declare the value type of a function explicitly. The reserved word **number** is used to denote the numeric value type. The format of a function type declaration is the same as for parameter and object types. That is, we could have written the functions above

```
(:functions
  (demand_chilled_goods ?c - location) - number ; remaining demand at ?c for
```

```
(demand_ambient_goods ?c - location) - number ; chilled/ambient goods
(free_capacity ?t - truck) - number            ; unused capacity in truck
(distance ?l1 ?l2 - location) - number          ; distance between locations
(per_km_cost ?t - truck) - number              ; per-kilometer cost of truck
(total-cost) - number
)
```

instead. Note that **number** can not be used as the type of any predicate or action parameters. This is one of the fundamental restrictions of numeric planning as defined in PDDL, as it ensures that the set of ground function terms, predicates, and actions remains finite.

Numeric Conditions and Effects

PDDL has five different types of numeric effects, but they all have a common form: (`<effect-type> <function-term> <expression>`), where the `<effect-type>` is one of `assign`, `increase`, `decrease`, `scale-up`, and `scale-down`. The `<function-term>` is formed by a function name with arguments drawn from constants and action parameters, in other words, a schema for the ground function term that will change when the effect takes place. The `assign` operator simply assigns the ground function term the value of the expression. The other four operators express a relative change: `increase` and `decrease` are additive changes, meaning the value of the expression is added to or subtracted from the ground function term, respectively, while `scale-up` and `scale-down` are multiplicative changes, meaning the ground function term is multiplied or divided by the value of the expression, respectively.

Numeric expressions are formed from function literals, constants and the four arithmetic operators (+, -, * and /). Constants can be integers or rational numbers written in decimal form. The operators take two arguments, although - can also be used as a unary operator. Some planners may allow + and * to take more than two arguments, although this form is nonstandard. Like all other expressions in PDDL, arithmetic expressions are written in parentheses and in prefix form, that is, with the operator first, then its arguments.

An atomic numeric condition is formed from one of the comparison operators =, <, <=, > or >= and two expressions. These are also written in parentheses and in prefix order. In the linehaul domain, as formulated in Section 2.1.3, the goal of the problem includes delivering all customer demands. In the numeric formulation, using the functions declared above, this can be written

```
(:goal (and (<= (demand_chilled_goods GV) 0)
            (<= (demand_ambient_goods GV) 0)
            (<= (demand_chilled_goods E) 0)
            (<= (demand_ambient_goods E) 0)
            (<= (demand_chilled_goods BW) 0)
            (<= (demand_ambient_goods BW) 0)
```

```
                (at ADoubleRef depot)
                (at BDouble depot)))
```

The use of <= instead of = allows for the goal to be achieved also if the plan delivers more than a customer asked for.

In our classical planning formulation of the linehaul domain, we used actions `deliver_ambient` and `deliver_chilled` to represent the delivery of one unit of ambient-temperature and chilled goods, respectively, by a truck at a customer location. In our numeric formulation, the first of these actions is written as follows:

```
(:action deliver_ambient
 :parameters (?t - truck ?l - location)
 :precondition (and (at ?t ?l)
                    (>= (free_capacity ?t) 1)
                    (>= (demand_ambient_goods ?l) 1))
 :effect (and (decrease (demand_ambient_goods ?l) 1)
              (decrease (free_capacity ?t) 1))
 )
```

The preconditions (>= (free_capacity ?t) 1) and (>= (demand_ambient_goods ?l) 1) ensure that the capacity of the truck is not exceeded and that the customer is not supplied with more than they asked for. It may be tempting to write an action like the following:

```
(:action deliver_ambient
 :parameters (?t - truck ?l - location ?q - number)
 :precondition (and (at ?t ?l)
                    (>= (free_capacity ?t) ?q)
                    (>= (demand_ambient_goods ?l) ?q))
 :effect (and (decrease (demand_ambient_goods ?l) ?q)
              (decrease (free_capacity ?t) ?q))
 )
```

with intention of being able to deliver an arbitrary quantity of goods at once. This, however, is not permitted, since action parameters can not have the type number. The extension of numeric planning to actions with numeric parameters, called *control parameters*, is being addressed in current research [e.g., Savas et al., 2016], but it is as yet not covered by PDDL.

We can, however, model a number of other delivery actions. The following action fulfills all the remaining demand of the customer at once, provided the truck has sufficient unused capacity to do so:

```
(:action deliver_ambient_all
 :parameters (?t - truck ?l - location)
 :precondition (and (at ?t ?l)
```

```
                        (>= (free_capacity ?t) (demand_ambient_goods ?l)))
 :effect (and (assign (demand_ambient_goods ?l) 0)
              (decrease (free_capacity ?t) (demand_ambient_goods ?l)))
 )
```

The expressions involved in the right-hand sides of numeric effects are all evaluated in the state before any of the effects are applied, or, in other words, all effects of an action are simultaneous. Thus, the value of (demand_ambient_goods ?l) on the right-hand side of the effect (decrease (free_capacity ?t) (demand_ambient_goods ?l)) is not zero, but the value that the function literal has before any of the action's effects are applied. We can of course write an analogous action for the case when the customer's demand exceeds the truck's unused capacity:

```
(:action deliver_ambient_max
 :parameters (?t - truck ?l - location)
 :precondition (and (at ?t ?l)
                    (>= (demand_ambient_goods ?l) (free_capacity ?t)))
 :effect (and (decrease (demand_ambient_goods ?l) (free_capacity ?t))
              (assing (free_capacity ?t) 0))
 )
```

Since the effect of both actions can be expressed as delivering the smaller of the two quantities, (demand_ambient_goods ?l) and (free_capacity ?t), they could have been written as one, had PDDL included a numeric min operator. However, since the minimum is defined by two cases, the two actions can be combined into one by using conditional effects, introduced in Section 3.1:

```
(:action deliver_ambient
 :parameters (?t - truck ?l - location)
 :precondition (at ?t ?l)
 :effect (and
         (when (>= (free_capacity ?t) (demand_ambient_goods ?l))
           (and (assign (demand_ambient_goods ?l) 0)
                (decrease (free_capacity ?t) (demand_ambient_goods ?l))))
         (when (< (free_capacity ?t) (demand_ambient_goods ?l))
           (and (assign (free_capacity ?t) 0)
                (decrease (demand_ambient_goods ?l) (free_capacity ?t)))))
 )
```

What if we want to model problems in which the customer's demands are non-integer quantities? The two actions above may be able to deal with this situation, but it does not allow in plans the choice of delivering an arbitrary amount from each visiting truck. As we have already seen in several examples, the expressive power of PDDL arises from the ability encode complex state transitions into sequences of actions. While we cannot write an action that takes an arbitrary numeric parameter, as shown above, we can write an action that instead takes the current value of another function as the quantity, and using a sequence of actions we can set the value of this function to any rational number that we want. Here is a possible formulation:

```
(:action drive
 ... ; as before
 :effect (and (assign (quantity) 1)
              ...) ; as before
 )


(:action deliver_ambient
 :parameters (?t - truck ?l - location)
 :precondition (and (at ?t ?l)
                    (>= (free_capacity ?t) (quantity))
                    (>= (demand_ambient_goods ?l) (quantity)))
 :effect (and (decrease (demand_ambient_goods ?l) (quantity))
              (decrease (free_capacity ?t) (quantity)))
 )


(:action scale_down_quantity
 :parameters ()
 :effect (assign (quantity) (/ (quantity) 10))
 )
```

The ground function term (quantity) is the quantity that will be delivered when deliver_ambient action is applied. We modify the drive action to include the effect (assign (quantity) 1), so that the value of (quantity) is 1 when a truck arrives at a customer. The action scale_down_quantity divides the current value of the function by 10, so that the next delivery action adds a fraction of the next decimal. Thus, delivery of 2.31 units of goods to customer E from truck BDouble is achieved by the following plan:

```
(drive BDouble depot E)     ; (quantity) becomes 1
(deliver_ambient BDouble E) ; deliver 1 unit
(deliver_ambient BDouble E) ; deliver 1 unit
(scale_down_quantity)       ; (quantity) becomes 1/10
(deliver_ambient BDouble E) ; deliver 0.1 unit
(deliver_ambient BDouble E) ; deliver 0.1 unit
(deliver_ambient BDouble E) ; deliver 0.1 unit
(scale_down_quantity)       ; (quantity) becomes 0.1/10
(deliver_ambient BDouble E) ; deliver 0.01 unit
```

Numeric Facts

The initial state of the problem specifies the initial values of all relevant ground functions terms. This is done in the `:init` section of the problem definition, by adding facts of the form `(= <ground function term> <decimal-value>)`. The ground function term is composed of a function name and arguments, which are objects declared in the domain or problem. The initial value of a function is written as a finite decimal number. Thus, it must be a rational number. The initial state of the numeric PDDL formulation of our example linehaul problem (depicted in Figure 2.2 on page 25) is as follows:

```
(:init
  ;; initial location and capacity of trucks:
  (at ADoubleRef depot)
  (at BDouble depot)
  (= (free_capacity ADoubleRef) 40)
  (= (free_capacity BDouble) 34)
  ;; customer requests:
  (= (demand_chilled_goods GV) 18)
  (= (demand_ambient_goods GV) 12)
  (= (demand_chilled_goods E) 7)
  (= (demand_ambient_goods E) 2)
  (= (demand_chilled_goods BW) 3)
  (= (demand_ambient_goods BW) 0)
  ;; distances between locations:
  (= (distance depot GV) 573)
  (= (distance depot E) 896)
  (= (distance depot BW) 876)
  ...
  (= (distance BW E) 79)
  ;; per-kilometer cost for the trucks:
  (= (per_km_cost ADoubleRef) 3.04)
  (= (per_km_cost BDouble) 2.59)
  ;; initialise total cost to zero:
  (= (total-cost) 0)
)
```

Logical and numeric facts can appear in any order. Each ground function term can only appear in one numeric fact in the `:init` section.

A ground function term that is not given a value in the initial state of the problem is *undefined*. The plan validity condition for numeric PDDL includes that every ground function term that appears in a precondition, right-hand side expression of a numeric effect, or as the target (left-hand side) of a relative numeric change effect of any ground action in the plan must be defined in the state that the action is applied to. Likewise, all ground function terms that appear in the goal must be defined in the final state. The only thing that a plan may do to an undefined function term is to assign it a value, with an `assign` numeric effect; this makes the function defined in the following state. We will define this more formally in Section 4.2.

The Plan Metric

The numeric fragment of PDDL allows for a very flexible specification of the objective function, referred to as the *plan metric* to be optimised as part of the planning problem definition. As we mentioned in Chapter 2, the way that action costs are specified in PDDL is a special case of a plan metric. The general form of a metric specification is

```
(:metric <direction> <expression>)
```

where `<direction>` is one of the keywords `minimize` or `maximize`, and `<expression>` is any ground numeric expression. The `:metric` specification is placed in the problem definition, not the domain, so that the expression can refer to objects declared in the problem. The metric value of a plan is the value of the expression in the final state reached by the plan's execution. In the case of sum of action costs, the value of the `(total-cost)` function is initially set to zero, and each action increases it by the cost of the action; thus, its value at the end of plan execution is the sum of the costs of the actions in the plan.

In the numeric formulation of the linehaul problem, we have used the same objective, `minimize (total-cost)`, as in the classical formulation in Chapter 2. However, to illustrate some of the range of plan metrics that are possible in numeric PDDL, we consider some variations on the problem. Suppose, for example, that total customer demands exceed the capacity of the whole truck fleet, so that there is no plan that achieves the goal. We could then relax the requirement that all demands are met, making the goal only that all trucks complete their routes (return to the depot) and the metric to minimise the total remaining demand:

```
(:metric minimize
         (+ (demand_chilled_goods GV)
            (+ (demand_ambient_goods GV)
               (+ (demand_chilled_goods E)
                  (+ (demand_ambient_goods E)
                     (+ (demand_chilled_goods BW)
                        (demand_ambient_goods BW)))))))
```

We may even express a trade-off between the cost of driving the trucks, recorded in `(total-cost)`, and the price paid by customers for the goods delivered:

```
(:metric maximize
         (- (+ (* (- 18 (demand_chilled_goods GV))
                  (price_chilled_goods GV))
```

```
(+ (* (- 12 (demand_ambient_goods GV))
      (price_ambient_goods GV))
   (+ ...
   ... ))) ; etc for all customers
(total-cost)))
```

It is worth keeping in mind, however, that although we can specify problems with complex plan metrics in PDDL, not many current planners are able to effectively optimise them; the majority can, at best, minimise the sum of constant action costs. This is an area where further planning research is needed.

4.2   NUMERIC PLAN VALIDITY

In the classical fragment of PDDL a plan, as defined in Section 2.2, is a sequence of ground action names, and this remains the case for the numeric planning subset of PDDL as well. The difference is that in numeric planning a state is no longer only a set of ground facts, but includes facts relating to the numeric functions. Unlike ground facts, which are always true or false, a ground function term can take any value in its numeric domain (which we take to be the rational numbers, though see discussion later in this chapter), or no value at all. In the latter case, the ground function term is said to be *undefined*. PDDL allows functions to be undefined in the initial state, and to remain undefined through part or all of a valid plan's execution, but no expression or condition that occurs in a ground action in the plan may be undefined. The only valid occurrence of an undefined ground function term in a plan is as the left-hand side (i.e., target) of an `assign` effect.

Let D be a domain definition in the numeric subset of PDDL and P a problem definition for D. We write functions(D) for the set of functions defined in D. As with actions and predicates, name(f) and param(f) denote the name and parameters, respectively, of each function $f \in$ functions(D).

Definition 4.1   Let $f \in$ functions(D) be a function, with param(f) $= x_1, \ldots, x_k$. Let $o_1, \ldots, o_k$ be a sequence of object names such that $o_i \in$ objects(P) and o_i IS-A type-of(x_i) for $i \in 1, \ldots, k$. Then (name(f) $o_1 \ldots o_k$) is a *ground function term*.

A ground function term is sometimes also called a (ground) *fluent*, or a *primitive numeric expression* [PNE; Fox and Long [2003]].

As in the case of classical planning, a plan is a sequence of (type-correct) ground action names (Definitions 2.2, 2.4), where each action name n corresponds to a grounding of an action schema a in the domain via a substitution σ that maps each parameter of the schema to the corresponding argument (object). From this, we obtain pre(n), add(n), and del(n) as the ground precondition formula, positive effects and negative effects of n, respectively (Definition 2.7). For the numeric planning case, we extend this definition to include that NE(a) is the set of ground numeric effects that appear in the formula $\sigma(\text{effect}(a))$. Each effect in $\text{NE}(a)$ is of the form (`<effect-type>` f e), where the effect type is one of the operators `assign`, `increase`, `decrease`, `scale-up`, or `scale-down`, f is a ground function term and e a ground numeric expression.

The semantics of sequential numeric plans are intuitively a straightforward extension of that of classical plans. The plan induces a state sequence, where the first state is the initial state of the problem and each following state is defined by applying the next action's effects in the previous state. Defining it precisely though is somewhat complex, due to the need to treat predicates and functions separately, and the conditions necessary to ensure that numeric expressions are well-defined.

Definition 4.2   Let F be the set of all ground facts and X the set of all ground function terms. A *state* is a pair $s = (s^F, s^X)$ where s^F is a subset of F and s^X is a partial mapping from X to $\mathbb{Q}$, the set of rational numbers.

As in the case of classical planning, the value of a ground compound numeric expression in a state is defined recursively in the usual manner. The same is true with a comparison operator applied to numeric arguments and a ground formula, which may combine numeric and predicate atoms.

Thus, with slight abuse of notation we will write $s(e)$ for the value of a ground numeric expression e in $s = (s^F, s^X)$, and $s(\varphi)$ for the value of any formula in s. However, since s^X is permitted to be a partial function, not every numeric expression or formula is necessarily defined in a state.

Definition 4.3   Let $s = (s^F, s^X)$ be a state. A ground function term f is *defined in s* iff s^X assigns f a value, i.e., $s^X(f)$ is defined. A ground numeric expression or comparison is defined in s iff all of its arguments are defined. A ground fact is always defined.

Somewhat surprisingly, there is no consensus on the exact definition of when a non-atomic formula (i.e., a formula composed of negation, conjunction and/or disjunction) should be considered defined in a state. It is easy enough for conjunctions: an undefined atomic formula cannot be considered true, and hence a conjunction containing undefined parts should also be false. Negations are trickier: if (x) being undefined means that (> (x) 0) is "false" (not satisfied because it is undefined), then perhaps (not (> (x) 0)) should be true, but on the other hand (<= (x) 0) is also undefined, and therefore not true. As long as negations are applied only to atomic formulas (i.e., predicates or comparisons between numeric expressions), we can fairly safely assume they are false when the negated formula is undefined. The most difficult is disjunction: if (x) is positive and (y) is undefined, is (or (> (x) 0) (> (y) 0)) true, or undefined? An argument can be made for either. The article by Fox and Long [2003] that introduced the numeric version of PDDL does not provide any answer, and different implementations of planners and plan validators have taken different interpretations. A problem modeller's best course of action may just be to avoid using disjunctions over numeric conditions that may be undefined, for example by translating them into different actions using the method sketched in Section 3.5.

Recall that each ground numeric function term can appear in at most one fact in init(P); a problem definition that specifies more than one initial value for any ground function term is invalid. The same is true also if a ground action's numeric effects specify potentially conflicting values for some ground function term.

Definition 4.4   Let n be a ground action. The set of numeric effects NE(n) is *non-conflicting* iff for every ground function term f, either there is at most one effect on f in NE(n) or all effects on f are additive (i.e., use the `increase` or `decrease` operators).

A plan is invalid if any ground action in the plan has conflicting numeric effects. Note that this is the case even if, in a particular state, the action's effects assign non-conflicting values: simply the potential for ambiguity about the result of applying the action rules it out from being part of a valid plan. Arguably, if an action has only multiplicative effects on a ground function term f, this could also be considered non-conflicting, since the combined effect is unambiguous. However, the article by Fox and Long [2003] that introduced the numeric subset of PDDL defines only additive effects to be non-conflicting, so it is safer to assume that planners, and plan validators, will treat multiplicative effects as conflicting.

Definition 4.5   Let $n_1, \ldots, n_l$ be a plan. The *induced state sequence* $s_0, \ldots, s_l$ is defined inductively as follows.

The initial state is $s_0 = (s_0^F, s_0^X)$, where s_0^F is the set of a ground facts in init(P), $s_0^X(f) = v$ if there is a ground numeric fact (= f v) in init(P) and $s_0^X(f)$ is undefined otherwise.

Let $s_{i-1} = (s_{i-1}^F, s_{i-1}^X)$: $s_i = (s_i^F, s_i^X)$, where,

- $s_i^F = (s_{i-1}^F \setminus \mathrm{del}(n_i)) \cup \mathrm{add}(n_i)$.
- $s_i^X(f) = s_{i-1}^X(f) + \sum_{e \in E^+} s_{i-1}^X(e) - \sum_{e \in E^-} s_{i-1}^X(e)$ if $E^+ \cup E^- \neq \emptyset$, where E^+ is the set of ground numeric expressions e such that (`increase` f e) $\in$ NE(n_i) and E^- is the set of ground numeric expressions e such that (`decrease` f e) $\in$ NE(n_i).
- $s_i^X(f) = s_{i-1}^X(e)$ iff (`assign` f e) $\in$ NE(n_i).
- $s_i^X(f) = s_{i-1}^X(f) \times s_{i-1}^X(e)$ iff (`scale-up` f e) $\in$ NE(n_i).
- $s_i^X(f) = s_{i-1}^X(f) \times (1/s_{i-1}^X(e))$ iff (`scale-down` f e) $\in$ NE(n_i) and $s_{i-1}^X(e) \neq 0$.
- $s_i^X(f) = s_{i-1}^X(f)$ if there is no effect on f in NE(n_i).

Again, note that if the problem definition is valid and the actions in the plan are all non-conflicting, then the induced state sequence is uniquely defined.

Definition 4.6   Let $n_1, \ldots, n_l$ be a plan and $s_0, \ldots, s_l$ its induced state sequence. The plan is *executable* iff (*i*) pre(n_i) is both defined and satisfied in s_{i-1}; (*ii*) each numeric expression appearing on the right-hand side of any effect in $\mathrm{NE}(n_i)$ or as the target of any relative effect in $\mathrm{NE}(n_i)$ is defined in s_{i-1}; and (iii) n_i is non-conflicting, for $i = 1, \ldots, l$. The plan is *valid* iff it is executable and goal(P) is both defined and satisfied in s_l.

4.3 MORE MODELLING EXAMPLES

In this section, we present a complex example of modelling numeric planning problems in PDDL. As in Chapter 2, the example helps to illustrate how we can work around some of the limitations of the language and make the most of its expressivity.

Racetrack is a game, traditionally played with pen on a squared paper [Gardner, 1973]. A car, represented by a point on the grid, moves with a velocity in each dimension (x and y). At each move, the car can adjust the velocity by ± 1 in one of the dimensions. (The game usually allows for changing both at once; we will model only the case of changing one for simplicity.) The goal is to drive the car around a track, defined by "walls" drawn on the paper, to a finish line, without ever crossing a wall. The game is usually played with two players, each controlling one car, and the winner is the one who reaches a finish line first. Here, we will model a situation with only one car whose goal is to reach a certain area. (In the two-player game, the goal is often to complete a cricuit around a track, which is also possible to encode by using several subgoals, but it more complicated. The two cars must also avoid running into one another.)

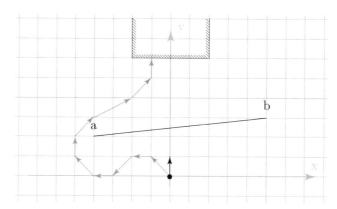

Figure 4.1: A simple example of the racetrack game, and a sample solution. The black point and arrow represent the starting position and velocity of the car. The goal is to avoid the obstacle and reach the centre region at the top, with no velocity in the x direction. The sequence of grey arrows shows a possible path.

A simple example is shown in Figure 4.1. In this example, there is only one wall. The car is initially at $x = 0, y = 0$, with a velocity $v_x = 0, v_y = 1$. The goal is to avoid the wall and reach the region in the centre above, with $v_x = 0$.

Modelling the movement of the car is simple: two functions (x) and (y) represent the position and two functions (vx) and (vy) represent its velocity. For example, the action that increases (vy), is defined as

```
(:action drive_plus_y
 :parameters ()
 :precondition ... ; see below
 :effect (and (increase (x) (vx))
              (increase (y) (+ (vy) 1))
              (increase (vy) 1))
)
```

The more difficult part is how to ensure the car does not cross a wall. Each move is a straight line segment, in the example above from x, y to $x + v_x, y + v_y + 1$. A wall is also a line segment, and can be represented by its two end points $a = (a_x, a_y)$ and $b = (b_x, b_y)$. (We assume $b_x > a_x$.) Each line segment is part of an infinite line, which can be represented by the equation $y = dx + c$. (This assumes the line is not vertical; we will deal with that situation separately below.) The slope of the movement direction is $d_{move} = \frac{v_y + 1}{v_x}$, and the slope of the wall is $d_{wall} = \frac{b_y - a_y}{b_x - a_x}$. The respective intercepts, c_{move} and c_{wall}, can be obtained from either end point of the line segment through $c = y - dx$. If the move and the wall are parallel, they will cross only if the intercepts are the same, and the two line segments overlap. The two lines are parallel iff $d_{move} = d_{wall}$, which is equivalent to $(v_y + 1)(b_x - a_x) = (b_y - a_y)v_x$. If the infinite lines are not parallel and neither of them is vertical, they will intersect at one point $x' = \frac{c_{wall} - c_{move}}{d_{move} - d_{wall}}$. In this case, the finite line segments cross iff $\min(x, x + v_x) \le x' \le \max(x, x + v_x)$ (remember that v_x can be negative) and $a_x \le x' \le b_x$. Finally, if the move is vertical (i.e., $v_x = 0$), a crossing can only occur at the car's current position in the x-direction. We obtain the position of the wall at this point in the y-direction as $y' = d_{wall}x + c_{wall}$, and the finite line segments cross iff $a_x \le x \le b_x$ and $\min(y, y + v_y + 1) \le y' \le \max(y, y + v_y + 1)$ (again, $v_y + 1$ can be negative). There is a similar special case if the wall is vertical. This means that the condition for the line segments to *not* cross is a disjunction, which must be repeated for each wall segment. A direct formulation results in a precondition like the following for the example action:

```
(:action drive_plus_y
 :parameters ()
 :precondition (forall (?w - wall)
               (or ;; The move and wall do not cross if
               ;; (i) the wall and the move are parallel, but either
               ;; they are not co-linear,
               (and (= (* (+ (vy) 1) (- (bx ?w) (ax ?w)))
                       (* (- (by ?w) (ay ?w)) (vx)))
                    (not (= (- (y) (* (/ (+ (vy) 1) (vx)) (x))) (c ?w))))
```

```
;; or they are not overlapping,
(and (= (* (+ (vy) 1) (- (bx ?w) (ax ?w)))
        (* (- (by ?w) (ay ?w)) (vx)))
     (< (x) (ax ?w))
     (< (+ (x) (vx)) (ax ?w)))
(and (= (* (+ (vy) 1) (- (bx ?w) (ax ?w)))
        (* (- (by ?w) (ay ?w)) (vx)))
     (> (x) (bx ?w))
     (> (+ (x) (vx)) (bx ?w)))
;; or (ii) the intercept point is outside one of the
;; line segments.
;; If neither move nor wall is vertical, then
;; xp = (/ (+ (- (c ?w) y) (* (x) (/ (+ (vy) 1) (vx))))
;;         (- (/ (+ (vy) 1) (vx)) (d ?w)))
;; and we need one of: x > xp and x + vx > xp,
(and (not (= (vx) 0))
     (> (x)
        (/ (+ (- (c ?w) (y)) (* (x) (/ (+ (vy) 1) (vx))))
           (- (/ (+ (vy) 1) (vx)) (d ?w))))
     (> (+ (x) (vx))
        (/ (+ (- (c ?w) (y)) (* (x) (/ (+ (vy) 1) (vx))))
           (- (/ (+ (vy) 1) (vx)) (d ?w)))))
;; x < xp and x + vx < xp,
(and (not (= (vx) 0))
     (< (x)
        (/ (+ (- (c ?w) (y)) (* (x) (/ (+ (vy) 1) (vx))))
           (- (/ (+ (vy) 1) (vx)) (d ?w))))
     (< (+ (x) (+ (vx) 1))
        (/ (+ (- (c ?w) (y)) (* (x) (/ (+ (vy) 1) (vx))))
           (- (/ (+ (vy) 1) (vx)) (d ?w)))))
;; ax > xp,
(and (not (= (vx) 0))
     (> (ax ?w)
        (/ (+ (- (c ?w) (y)) (* (x) (/ (+ (vy) 1) (vx))))
           (- (/ (+ (vy) 1) (vx)) (d ?w)))))
;; or bx < xp,
(and (not (= (vx) 0))
     (< (bx ?w)
        (/ (+ (- (c ?w) (y)) (* (x) (/ (+ (vy) 1) (vx))))
```

```
                    (- (/ (+ (vy) 1) (vx)) (c ?w)))))
        ;; since we assume that ax < bx.
        ;; If the move is vertical, we calculate
        ;; yp = (+ (* (d ?w) (x)) + (c ?w)), and check that
        ;; y > yp and y + vy + 1 > yp,
        (and (= (vx) 0)
             (> (y) (+ (* (d ?w) (x)) + (c ?w)))
             (> (+ (y) (+ (vy) 1)) (+ (* (d ?w) (x)) + (c ?w))))
        ;; y < yp and y + vy + 1 < yp,
        (and (= (vx) 0)
             (< (y) (+ (* (d ?w) (x)) + (c ?w)))
             (< (+ (y) (+ (vy) 1)) (+ (* (d ?w) (x)) + (c ?w))))
        ;; ax > x, or
        (and (= (vx) 0) (> (ax ?w) (x)))
        ;; bx < x, or
        (and (= (vx) 0) (< (bx ?w) (x)))
        ))
 :effect (and (increase (x) (vx))
              (increase (y) (+ (vy) 1))
              (increase (vy) 1))
 )
```

We have left out the case when the wall is vertical, and we have introduced two extra static functions, (d ?w) and (c ?w), for the slope and intercept of each wall ?w, instead of substituting the expressions needed to calculate them from the end point coordinates. Even so, we can note that several complex expressions appear repeatedly in different places in the precondition. This is because numeric PDDL does not have a mechanism for defining "abbreviations", such as local variables in action definitions or an analogue of axioms for derived functions.

Several of the expressions in the precondition above can potentially result in a division by zero. As noted in Section 4.2, there is some ambiguity about whether a disjunctive condition is satisfied in a state where some of its disjuncts are undefined while others are true. Likewise, it has not been clearly established whether division by zero makes a numeric expression undefined, or if it should be evaluated to infinity. (As can be seen from the example action above, the latter interpretation leads to further questions, such as whether infinity divided by infinity is undefined, or if not, what its value is.) Consequently, whether the action definition above works as intended may differ between different planners.

To guard against ambiguity, we can rewrite the domain model to use a sequence of actions to check for collisions with each wall in turn, which allows us to break up the cases of the disjunction into preconditions of separate actions (like the compilation described in Section 3.5). Since in every case where a move does not cross a wall, at least one disjunct will be both defined and true, at least one of the corresponding actions will be applicable. Another advantage of such a formulation is that the set of planners that support both numeric PDDL and complex preconditions (involving `forall` and `or`) is very limited.

4.4 COMPLEXITY OF NUMERIC PLANNING

As discussed in Section 2.5, we distinguish several computational problems related to a planning problem: plan validation, plan existence, plan generation (or proof of non-existence), and plan cost optimisation. The numeric extension of PDDL does not change the complexity of plan validation, which is still tractable, but, without further restriction on the use of numeric functions in the domain, the other questions become undecidable. This is because unrestricted counting allows for problems that have no bound on plan length. Plan existence and generation are semi-decidable because we can still enumerate action sequences in order of increasing length and stop with a positive answer if a valid plan is found, but without an upper bound on plan length we cannot prove plan non-existence by enumeration.

In the following sections, we will sketch the proofs of undecidability and decidability of a few restricted subsets of numeric planning. Both kinds of results are useful for understanding more precisely where the difficulty in numeric planning lies.

4.4.1 SOME UNDECIDABLE CASES

We present two reductions from known undecidable problems to plan existence for the numeric fragment of PDDL. Each reduction uses a different, but in both cases quite restricted, subset of the language, but both of them use only ground PDDL. Both were shown by Helmert [2002].

The first reduction is from the problem of finding an integer solution to a Diophantine equation, which is an equation of the form $p(x_1, \ldots, x_n) = 0$ where $p(x_1, \ldots, x_n) = 0$ is a polynomial in multiple variables, $x_1, \ldots, x_n$. This is an undecidable problem.

To construct a planning problem from the equation $p(x_1, \ldots, x_n) = 0$, we use n functions, $(x_1), \ldots, (x_n)$, and pose $p((x_1), \ldots, (x_n)) = 0$ as the goal. We then define actions which allow the functions $(x_1), \ldots, (x_n)$ to be set to any integer value: if an integer solution to the equation exists, a plan exists by setting the values of the functions to it; conversely, if a plan exists then the values of the functions in the final state reached by executing this plan must be an integer solution to the equation. To allow the functions to take any integer value, we only have to set them to 0 in the initial state and define for each variable one action that increments it by 1 and one that decrements it by 1.

This construction uses only additive effects where the right-hand side of each `increase` and `decrease` effect is a constant, and it has no numeric action preconditions. The language subset used by the construction can be further restricted by modelling each variable (x_i) with two variables, (pos_i) and (neg_i), and replacing (x_i) in the goal condition with the difference (- (pos_i) (neg_i)). Instead of actions increasing and decreasing (x_i), we then have only actions that increase (by a constant of 1) (pos_i) and (neg_i). This makes the domain "monotonic", in the sense that all variables only change in one direction, and the state space acyclic. However, the construction still needs a complex goal condition.

The second reduction is from the halting problem of an *abacus program*, also known as a *counter machine* [Boolos and Jeffrey, 1980]. Informally, this is a finite state machine augmented with a fixed set of integer counters. Machine state transitions can increase a counter by one, or decrease a counter and conditionally jump to different states depending on if the counter has reached zero or not. The problem of deciding if a deterministic abacus program when executed from a given starting configuration will ever reach a designated halting state or not is undecidable, if the program uses more than one counter.

Modelling the discrete state mechanics of the program in PDDL is straightforward, similar to how, for example, we modelled the dining philosophers (which are also finite state machines) in Section 2.4.2. A step that changes the program state from l to m, and increments counter x, is modelled with an action like the following:

```
(:action step_{l,m}
 :precondition (at_l)
 :effect (and (not (at_l))
              (at_m)
              (increase (x) 1))
 )
```

A step that decrements a counter is modelled with two actions, since the next state depends on the value of the counter. For example, if the destination is m if the counter x reaches zero and k otherwise, we get the two actions

```
(:action step-if-zero_{l,m}
 :precondition (and (at_l) (= (x) 1))
 :effect (and (not (at_l))
              (at_m)
              (decrease (x) 1))
 )
```

```
(:action step-if-non-zero_{l,k}
 :precondition (and (at_l) (> (x) 1))
 :effect (and (not (at_l))
              (at_k)
              (decrease (x) 1))
 )
```

It is not hard to show that this construction allows only for plans that emulate the actions of the abacus program. Setting the initial state as the starting configuration of the program and posing as the goal (at_h) for the designated halting location h completes the reduction.

This construction differs from the previous in that all atomic numeric conditions and effects are "simple". Conditions are only comparisons between a single function and a constant, and effects again are only of the form increase or decrease by a constant. In this example, the complexity arises from the interaction between effects and conditions, in particular the fact that it uses two kinds of comparisons that are each others complement ($= 0$ and > 0).

4.4.2   SOME DECIDABLE CASES

A case of numeric planning which is easy to show decidable is if all numeric effects have the form of assigning a constant value. Since the number of such effects that can appear in the domain is fixed, there is only a finite set of values that numeric functions can take. The possible numeric literals $(= f \ v)$ can be replaced with a set of predicates, though this must be done in a way that takes into account the initial state specification of the problem as well, and all numeric conditions over those values replaced with formulas over the new predicates. The problem is essentially classical.

As an example of a case that is decidable in a very non-trivial way, we can take domains in which all numeric effects are constant additive effects and the only numeric preconditions is that every action that has an effect of the form (decrease f c) also has the precondition (>= f c), preventing any numeric function from becoming negative. If the goal is a conjunction of equalities and inequalities between numeric functions and constants, plan existence corresponds precisely with the reachability problem of the Petri net formalism, which decidable [Mayr, 1981]. (The propositional aspects of the domain can be modelled with bounded places in the Petri net. Petri nets only allow for integer constants, but since the set of rational constants that appear in the domain and problem is finite, they can be converted to integers by finding their common denominator.) This example shows an interesting contrast with the reduction from abacus programs outlined in the previous section, in that that construction nearly falls into this decidable fragment, but for the presence of equalities between numeric functions and constants in action preconditions.

A number of other restricted fragments of numeric planning were shown to be decidable by Helmert [2002].

4.4.3   NUMERIC PLANNING AND OTHER EXTENSIONS

Combining the numeric subset of PDDL and the extensions of the classical model described in Chapter 3 is straightforward. In this chapter, we have already used actions with quantified and disjunctive numeric preconditions; likewise, numeric effects can be conditional, and effect conditions can be numeric. Numeric functions can not be defined by axioms. That is, writing an axiom such as

```
(:derived (= (d ?w - wall) (/ (- (by ?w) (ay ?w)) (- (bx ?w) (ay ?w)))))
```

is *not* permitted. Therefore, the combination of numeric planning with axioms does not alter the complexity of the problem. Numeric state conditions appearing in trajectory constraints or preferences also do not change how these are evaluated, or can be compiled away.

As will be seen in the following two chapters, the combination of temporal and numeric planning allows for defining problems with continuous change.

CHAPTER 5

Temporal Planning

In Chapter 4, we extended classical planning to variables that are numeric in nature. In this chapter, we extend classical planning in an orthogonal way, through the introduction of *time*. In temporal planning, actions are durative in nature, and both the conditions and effects of an action must be generalised accordingly. Temporal planning was introduced into PDDL with the 2002 IPC, although of course there were temporal planners long before that [e.g., Ghallab and Laruelle, 1994, Muscettola, 1994, Penberthy and Weld, 1994, Smith and Weld, 1999, Vere, 1983, to name just a few] and the majority of this chapter follows the definition of PDDL version 2.1 [Fox and Long, 2003].

Scheduling—roughly approximated, the task of deciding on *when* to do things—is an area of AI closely related to automated planning, and temporal planning builds on the combination of classical planning formalism and scheduling. A solution to a temporal planning problem is no longer simply a sequence of actions, but rather a timed schedule of (t_i, a_i) pairs where t_i is a time point and a_i is either the start or end of an action. As discussed in detail later in Section 5.3, the validity of a solution to a temporal planning problem relies not only on the sequence of states it yields (as is the case with classical planning), but also on the notion of the temporal consistency: each action's duration in the solution must respect the domain specification, which restricts the set of possible durations.

As an added convenience for modelling temporal problems, the temporal planning fragment of PDDL also includes *Timed Initial Literals* (TILs). These are facts in the problem that are known to become true or false after a preset amount of time (e.g., "the sun will rise at 8am" or "the machine is only available until 13:00"). As we shall see in Section 5.2, TILs provide a powerful modelling tool for domain designers to represent complex behaviour that marries temporal aspects of the domain and the state of the world.

We begin by introducing durative actions in Section 5.1. The key difference between actions in classical planning and temporal planning is that in the latter we must specify if an effect occurs at the start or end of an action as well as if a condition is for the start, end, or entire duration. Continuous change to a state variable over the duration of an action only makes sense when the state representation allows for a non-binary representation, and we address this important extension (i.e., the combination of numeric and temporal planning) in Section 5.4. Further, in Chapter 6 we will consider complex interactions between numeric state variables and action duration.

5.1   DURATIVE ACTIONS

Until now, we have presumed actions to be instantaneous occurrences that transition the world from one state to another. In its most basic form, temporal planning does not remove the instantaneous transition of one state to another (a topic left for hybrid planning in Chapter 6), however the conditions on actions and the manner in which they affect the world is generalised to address a span of time rather than an instantaneous occurrence. From the modelling perspective, we distinguish between a classical instantaneous action and a durative action by using (`:durative-action ...`) to describe the latter in the domain description.

The use of durative actions is indicated in a planning domain by adding `:durative-actions` to the list of `:requirements`. Further, if we wish to optimise for the total time a plan takes to achieve its goal, the following metric is added to the problem description:

```
(:metric minimize (total-time))
```

Note that the function term (`total-time`) is implicitly defined in temporal planning domains; it does not need to be declared in the `:functions` section, and it cannot be used anywhere else in the domain or problem.

A useful feature for defining the duration of actions is to condition it on the parameters of the action. This is done by defining (usually static) functions which are used to write expressions that evaluate to the duration, in the same way that we have seen expressions for action costs in Chapter 2 and general numeric expressions in Chapter 4. As an example, this is how we would define a function for the driving time between two locations in the domain description:

```
(:functions (drive_time ?l1 ?l2 - location))
```

To make use of this, the initial state section of the problem definition must include assertions such as (= (`drive_time Ottawa Montreal`) `120`). Note that the specific units used for the constants and durations are up to the domain designer's interpretation (in the previous example, this would be 120 minutes). That said, every duration specified is assumed to be in the same time unit, and may be any non-negative real number. Consider the following example of a durative action that includes all of the new components introduced for specifying the more complex action type:

```
(:durative-action drive
 :parameters (?c - car ?l1 ?l2 - loc)
 :duration (= ?duration (drive_time ?l1 ?l2))
 :condition (and (at start (car_at ?c ?l1))
                 (over all (road_open ?l1 ?l2))
                 (at end (free_space_at ?l2)))
 :effect (and (at start (not (car_at ?c ?l1)))
```

```
     (at start (free_space_at ?l1))
     (at end (car_at ?c ?l2))
     (at end (not (free_space_at ?l2)))))
)
```

The first thing we notice in this example is the duration section. The syntax is (= ?duration <expression>), where <expression> can be any numeric expression, built from constants, function terms and arithmetic operators, as we have seen in the preceding chapters. Here, the duration of the drive action is the driving time between the two locations that instantiate the durative action.

Next, we have the :condition section of the drive action (note the difference from a traditional action's :precondition section). The types of conditions fall into three main categories: (1) conditions that must hold when the action begins (specified as (at start ...)); (2) conditions that must hold during the entire action execution (specified as (over all ...)); and (3) conditions that must hold when the action completes (specified as (at end ...)).

We can see that in the drive action, the car must start at location ?l1. For the duration of the action, the road must be open as well—this may model a domain where roads can be closed for maintenance. Note that the road should also be open when the action begins, but it is more appropriately assigned to an (over all ...) condition, as we do not want a road to be closed off while cars are currently driving. Finally, the last condition stipulates that there is space at the destination location: for demonstrative purposes, we assume that each location can only hold a single car at a time. Note that the car can begin to drive when the destination location does not have space, as long as space becomes available prior to the car arriving.

Next, we have the effects of the action. As mentioned earlier, effects of a durative action are instantaneous and appear either at the start or end of the action (using the same syntax as the :condition section). At the start of the drive action, we both delete the fact that the car is at the source location and add the fact that there is now space there. Conversely, at the end of the action, we add the new location of the car and remove the fact that states there is space at the destination.

The drive action is depicted in Figure 5.1. The two instantaneous conditions appear before the start and end of the action (indicated by the vertical bars), and the over all condition is shown inside the action itself (as it must hold for the entire duration). The effects (both positive and negative) appear on the right side of the start and end action indicators.

Generalising to the temporal setting allows us to model the concurrent execution of overlapping actions. As a larger example, let's consider a temporal version of the elevator domain presented in Section 3.1.1. Example 13 below shows the beginning of a temporal elevator domain. We will show how to define the domain's actions later in this section.

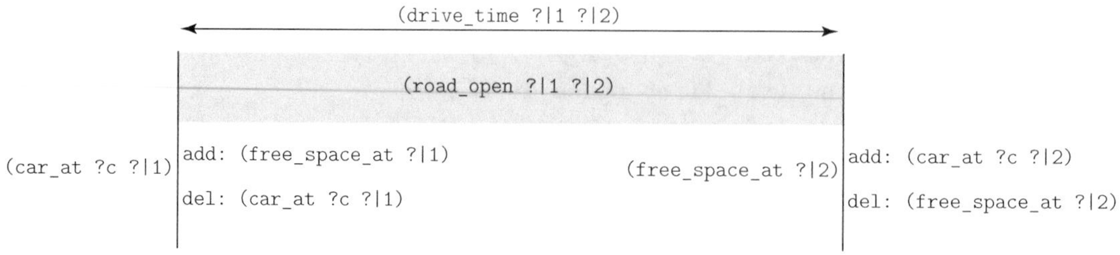

Figure 5.1: Visual representation of the drive action.

```
(define (domain temporal-elevators)
  (:requirements :typing :fluents :durative-actions)

  (:types elevator passenger num - object)

  (:predicates
   (passenger-at ?person - passenger ?floor - num)
   (boarded ?person - passenger ?lift - elevator)
   (lift-at ?lift - elevator ?floor - num)
   (next ?n1 - num ?n2 - num)
   )

  (:functions
   (person_speed ?person - passenger)
   (elevator_speed ?lift - elevator)
   (floor_distance ?f1 ?f2 - num)
   )

  ;; Actions...
  )
```

PDDL Example 13: Elevator domain using durative actions.

We now have functions defined to represent the individual passenger's speed, elevator speed, and distance between floors. We assume that the units for the numeric constants are consistent. E.g., meters/sec for speed, seconds for time, and meters for distance. Now consider the elevator movement actions:

```
(:durative-action move-up
 :parameters (?lift - elevator ?cur ?nxt - num)
 :duration (= ?duration (/ (floor_distance ?cur ?nxt)
                           (elevator_speed ?lift)))
 :condition (and (at start (lift-at ?lift ?cur))
                 (over all (next ?cur ?nxt)))
 :effect (and (at start (not (lift-at ?lift ?cur)))
              (at end (lift-at ?lift ?nxt)))
 )


(:durative-action move-down
 :parameters (?lift - elevator ?cur ?nxt - num)
 :duration (= ?duration (/ (floor_distance ?cur ?nxt)
                           (elevator_speed ?lift)))
 :condition (and (at start (lift-at ?lift ?cur))
                 (over all (next ?nxt ?cur)))
 :effect (and (at start (not (lift-at ?lift ?cur)))
              (at end (lift-at ?lift ?nxt)))
 )
```

Notice the combination of values that are used to derive the duration of elevator movement: the distance divided by the speed. The over all condition is an invariant, as we never change the value of the next predicate in this domain. The other conditions and effects stipulate that the elevator changes location, but notice that *during* a move action, the elevator will not have any lift-at fact true; this rules out its use for boarding or disembarking passengers during this time.

Next, we have the actions for boarding and disembarking the passengers:

```
(:durative-action board
 :parameters (?per - passenger ?flr - num ?lift - elevator)
 :duration (= ?duration (person_speed ?per))
 :condition (and (over all (lift-at ?lift ?flr))
                 (at start (passenger-at ?per ?flr)))
 :effect (and (at start (not (passenger-at ?per ?flr)))
              (at end (boarded ?per ?lift)))
 )


(:durative-action leave
 :parameters (?per - passenger ?flr - num ?lift - elevator)
 :duration (= ?duration (person_speed ?per))
 :condition (and (over all (lift-at ?lift ?flr))
```

```
                    (at start (boarded ?per ?lift)))
  :effect (and (at end (passenger-at ?per ?flr))
              (at start (not (boarded ?per ?lift)))))
  )
```

The passengers themselves each have a speed at which they can enter the elevator, and we require that the elevator remain at the floor during this entire time. Similar to the lift location, all passenger-at facts for a particular passenger will be false during the execution of these actions. Finally, we have the problem definition extended to the temporal setting:

```
(define (problem elevators-problem)
  (:domain temporal-elevators)

  (:objects
  n1 n2 n3 n4 n5 - num
  p1 p2 p3 - passenger
  e1 e2 - elevator
  )

  (:init
  ;; Same fluents as the classical planning example
  (next n1 n2) (next n2 n3) (next n3 n4) (next n4 n5)
  (lift-at e1 n1) (lift-at e2 n5)
  (passenger-at p1 n2)
  (passenger-at p2 n2)
  (passenger-at p3 n4)

  ;; Define how fast each of the passengers move (in seconds)
  (= (person_speed p1) 2)
  (= (person_speed p2) 3)
  (= (person_speed p3) 2)

  ;; Define the speed of the elevators (in meters / second)
  (= (elevator_speed e1) 2)
  (= (elevator_speed e2) 3)

  ;; Define the distance between the floors (in meters)
  (= (floor_distance n1 n2) 3)
  (= (floor_distance n2 n3) 4)
  (= (floor_distance n3 n4) 4)
```

```
  (= (floor_distance n4 n5) 3)
  (= (floor_distance n5 n4) 3)
  (= (floor_distance n4 n3) 4)
  (= (floor_distance n3 n2) 4)
  (= (floor_distance n2 n1) 3)
  )

  (:goal (and (passenger-at p1 n1)
              (passenger-at p2 n1)
              (passenger-at p3 n1)
              ))

  (:metric minimize (total-time))
  )
```

PDDL Example 14: Elevator problem using durative actions.

___

The only differences in the temporal setting are the specification of the static function terms (used to define action duration) and the specification that we would like to minimize the total execution time taken of the plan. Figure 5.2 shows an example solution to the above problem (computed using an off-the-shelf temporal planner, Optic [Benton et al., 2012]). The domain can be found and interactively solved at `editor.planning.domains/pddl-book/ele` `vator`. The textual representation of the plan is given in Figure 5.3.

The individual colours represent different action types in the domain, and the x-axis corresponds to time. Note that many of the actions occur simultaneously: for example, boarding of elevator #2 begins while elevator #1 is still traveling, and both elevators move concurrently for a time.

Required Concurrency

In some cases, a temporal planning problem can be solved by modelling and solving it as a purely classical problem, ignoring time, and then scheduling the actions in the resulting plan to satisfy duration and precedence constraints. For example, closer inspection of the schedule in Figure 5.3 shows that each action simply starts (almost) as early as possible, and ends after the required duration. (For example, (board p1 n2 e1) starts as soon as (move-up e1 n1 n2) ends; it could not start earlier since the preceding action adds (lift-at e1 n2) which is required over all by the boarding action.) This method of temporal planning fails, however, if there are actions whose execution *must* overlap for a plan to be valid. This is known as *required concurrency*.

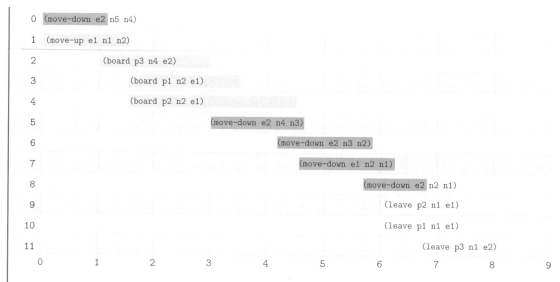

Figure 5.2: A visual representation of the plan for the temporal elevator problem.

```
0.000: (move-down e2 n5 n4)   [1.000]
0.000: (move-up e1 n1 n2)   [1.500]
1.000: (board p3 n4 e2)   [2.000]
1.500: (board p1 n2 e1)   [2.000]
1.501: (board p2 n2 e1)   [3.000]
3.000: (move-down e2 n4 n3)   [1.333]
4.334: (move-down e2 n3 n2)   [1.333]
4.501: (move-down e1 n2 n1)   [1.500]
5.668: (move-down e2 n2 n1)   [1.000]
6.001: (leave p2 n1 e1)   [3.000]
6.001: (leave p1 n1 e1)   [2.000]
6.668: (leave p3 n1 e2)   [2.000]
```

Figure 5.3: A plan for the temporal elevator problem (Example 14).

As an example, let us model the decision to open the elevator doors at a floor with an explicit action:

```
(:durative-action open-door
 :parameters (?lift - elevator)
 :duration (= ?duration (door_speed ?lift))
 :condition (at start (not (open ?lift)))
```

```
:effect (and (at start (open ?lift))
             (at end (not (open ?lift)))))
)
```

We then modify the board and leave actions to require that (over all (open ?lift)) holds. There is now required concurrency between the action that holds the elevator door open, and the movement of passengers: the open-door action essentially envelopes the passengers' movement actions. Only passengers who move fast enough to complete their boarding or leaving action within the duration of the open-door action can now enter or leave the elevator, i.e., (person_speed ?per) must be less than (door_speed ?lift).

Example: Forcing Action Non-Overlap

In the preceding example, actions board and leave are forced to overlap with open-door. This is achieved through a predicate that is added only during the duration of the enveloping action and required by the enveloped actions. A complementary example is to force actions to *not* overlap through the use of a *unary resource*. Consider the following modification of the open-door action:

```
(:durative-action open-door
 :parameters (?lift - elevator ?flr - num)
 :duration (= ?duration (door_speed ?lift))
 :condition (and (at start (not (open ?lift)))
                 (at start (can-stop-at ?flr))
                 (overall (lift-at ?lift ?flr)))
 :effect (and (at start (open ?lift))
              (at start (not (can-stop-at ?flr)))
              (at end (not (open ?lift)))
              (at end (can-stop-at ?flr)))
)
```

Here, the predicate can-stop-at restricts the domain such that only one elevator can open its door at a particular floor at any time. We achieve this by requiring the fact at the start of the action, and deleting it only for the duration of the action, i.e., deleting it at the start of the action and adding it again at the end. Note that this resource is associated with only the floor, and not the elevator as well.

Example: Restricting Plan Duration

As a final encoding trick for this section, we will introduce one further action to the elevators domain: execute. PDDL has no special-purpose mechanism for specifying a maximum plan duration, but we can get around this by creating an auxiliary action that envelopes every other action in the plan. The process is two-fold: (1) we add a predicate (enabled) to the domain and make it an `over all` condition for every action; and (2) we add the predicate (`can-execute`) to the domain and to the initial state, and add the following action to the domain (making the maximum duration 10 units of time):

```
(:durative-action execute
 :parameters ()
 :duration (= ?duration 10)
 :condition (at start (can-execute))
 :effect (and (at start (enabled))
              (at start (not (can-execute)))
              (at end (not (enabled)))))
)
```

Notice that because (`can-execute`) is both required and deleted at the start of `execute`, and not added by any action in the domain, the `execute` action can only occur once. Further, since (`enabled`) will only hold during that execution, and since every action has this fact as an overall condition, every action (aside from `execute`) must occur entirely within the duration of this new action. Changing the duration will change the maximum amount of time that is given for all other actions to occur.

Now that we have concluded a description of the basic temporal planning syntax and usage, we will move to describing a commonly used extension for modelling predictable events: timed initial literals.

5.2 PLANNING WITH PREDICTABLE EVENTS

Now that we have made the shift to a setting where actions must be orchestrated temporally in order to produce a solution, a natural question is how we may specify temporal restrictions on the state of the world. In the temporal fragment of PDDL, this can be done through what is known as *timed initial literals* (TILs): changes to the state of the world that occur a set time after the start of the plan.

TILs are specified as part of the initial state of a temporal problem, and may be either positive (adding a fact) or negative (deleting a fact). The syntax of TILs is:

```
(at <time> <literal>)
```

where `<time>` is a numeric constant (integer or decimal) and `<literal>` is a positive or negative ground literal. The status of the symbol `at` is somewhat ambiguous. Due to its special role in writing TILs, it could be considered a reserved word, at least in the temporal fragment of PDDL. However, there are also many examples of domain models that use `at` as a predicate, and many planners accept this dual use. The use of TILs in a problem is indicated with the keyword `:timed-initial-literals` in the `:requirements` section.

The introduction of TILs does not strictly increase the expressivity of temporal planning problems that can be defined, as one can use modelling tricks to capture the same behaviour, e.g., forcing a dedicated (short) action at the start of every valid plan, that enables a set of mandatory actions corresponding to each of the desired TILs. Actions can always be made mandatory by having them add a unique fact to the state of the world that also appears as a goal.

Despite the absence of additional expressivity, using TILs makes the encoding process far more natural: the initial state specification now includes what is true or false across time rather than just at the start of the plan. Some common uses of TILs include:

1. modelling deadlines for the planning to be complete;

2. providing a time window for certain actions to occur; and

3. forcing an action to occur at a specific time.

We will describe each of these use cases in turn. We direct the reader to Fox et al. [2004] for other practical temporal planning modelling tricks.

Example: Restricting Plan Duration

In the last section, we showed how to limit a plan's total execution time (also known as *makespan*) by creating an auxiliary action to envelope the entire plan. With TILs at our disposal, we can achieve this far more elegantly. Using the predicate introduced for restricting the plan duration in the elevator domain, we can forgo the extra `execute` action, and add the following to the initial state:

```
(enabled)
(at 10 (not (enabled)))
```

Similar to the previous encoding for restricting the plan duration, every action in the domain is predicated (using an `over all` constraint) on the fact (`enabled`). The new initial state restricts this fact to only hold between time $t = 0$ to $t = 10$.

Example: Enabling Action Windows

As a generalisation of the previous example, we may want to enable a *subset* of actions for a particular time period. For example, perhaps vehicles should only be driven between certain hours of the day, or particular resources may only be used at regular intervals. To model this using TILs, every set of actions we want to restrict for a time window must have an associated unique predicate such as (`enabled`) added as an `over all` condition, and the time windows $[t\_1\_start, t\_1\_end], [t\_2\_start, t\_2\_end], \ldots$ (where $t\_i\_start \leq t\_i\_end$) modelled as TILs:

```
(at t_1_start (enabled))
(at t_1_end (not (enabled)))

(at t_2_start (enabled))
```

```
(at t_2_end (not (enabled)))

   . . .
```

To ensure that this works as intended, the intervals must *not* overlap, which can be easily done by merging any pair of overlapping intervals into a single one.

Example: Forced Action in Time

In the previous example, time windows were specified allowing the possibility of actions to occur, but this by no means *forces* the actions to occur. If, instead, we would like to model the fact that a certain action *will* occur at a specific time, we can again use the power of TILs. Essentially, we combine the time window modelling technique above with the technique for forcing an action to appear in the plan. Assuming do-foo is the action we wish to force starting at a particular time t_1, and we are willing to be within a margin of error on the start time of ϵ such that $t_2 = \epsilon + t_1$, then we use the following TILs:

```
(at t₁ (do-foo-enabled))
(at t₂ (not (do-foo-enabled)))
```

Action do-foo must have (at start (do-foo-enabled)) in its :condition and (at end (do-foo-done)) in its :effect. Finally, adding (do-foo-done) to the problem's goal forces the action to occur in every plan, since it is the only one to add this fact, while the TILs force the action to occur only at a specific time.

While there is some room between t_1 and t_2 for the action timing, the difference between them can be controlled to within the precision allowable by the planner. We dive deeper into this key issue of exact timing in the next section as we discuss temporal plan validity.

5.3   TEMPORAL PLAN VALIDITY

We now turn to the semantics of temporal plans. What follows is a summary of the semantics of PDDL version 2.1 as defined by Fox and Long [2003], and we refer the reader to that paper for any details not covered here.

A solution to a temporal planning problem, as pointed out earlier in the chapter, is a collection of durative or instantaneous actions that are positioned in time. However, we will assume here that all actions in the plan are durative; this is only to simplify the presentation. An instantaneous action is semantically equivalent to a durative action whose duration is zero and whose conditions and effects are all placed at start. The semantics of a temporal plan are built on a notion called *happenings* which represent points in time where the state of the world changes due to the effects of one or more actions, or the occurrence of a TIL. Intuitively, these happenings make up a sequence of classical planning actions, and the validity of the temporal plan is thus reduced to validity of a classical plan, together with some conditions to ensure that it is well-defined.

Definition 5.1   A *temporal plan* is a multiset of action occurrences. Each action occurrence, t : n [d], consists of a start time, t, which is a non-negative rational number, a ground action name n (cf. Definition 2.1), and a duration d, which is also a non-negative rational number.

Each action's duration must be specified in the plan because PDDL allows durative actions to have constraints on the `?duration` parameter other than equality (see, for instance, Example 18 in Section 5.4 below). If the duration is not fixed to a single value, a planner can choose any value for the duration of each action occurrence in a plan that satisfies the duration constraints. The notation t : n [d] for action occurrences in a temporal plan is not defined as part of PDDL, but because it is the plan format required by the temporal plan validator VAL, it is a defacto standard.

A temporal plan is a multiset, rather than a set, of timed action occurrences because the semantics of temporal PDDL allow for multiple instances of an action to occur at the same time, in certain circumstances. The non-interference requirement (Definition 5.4) rules out most self-concurrent actions, but when we combine the temporal and numeric subsets of PDDL, simultaneous actions, including multiple simultaneous occurrences of the same action, can have a cumulative effect on numeric fluents that is the sum of the effect of each action occurrence. For example, executing k instances of the same action with the effect (`increase` f 1) at the same time results in the fluent f increasing by k.

The requirement that all ground action names in a plan match a schema in the planning domain and are instantiated with the right number of objects of the right types is the same as in the non-temporal case (cf. Definition 2.4). As before, assuming the plan is type correct, we identify each timed ground action occurrence with the corresponding grounded action definition from the domain. However, in the ground action corresponding to the action occurrence t : n [d], the `?duration` parameter is replaced with the constant d.

The execution of a temporal plan is defined via a corresponding *simple plan*, which is a multiset of time stamped instantaneous (STRIPS) actions. To define the simple plan, we must first define the instantaneous actions that are part of it.

Definition 5.2   Let a be a (ground) durative action. The instantaneous actions a_{start}, a_{end} and a_{overall} are defined such that:

- $\text{pre}(a_{\text{start}})$ is the set of literals appearing within the scope of the `at start` temporal specifier in the action's `:condition` or `:duration` formulas. Any atomic condition that appears without a temporal specifier in the `:duration` formula also belongs to $\text{pre}(a_{\text{start}})$.

- $\text{add}(a_{\text{start}})$, $\text{del}(a_{\text{start}})$ and $\text{NE}(a_{\text{start}})$ are the sets of added and deleted facts and numeric effects appearing within the scope of the `at start` temporal specifier in the action's `:effect`.

- $\text{pre}(a_{\text{end}})$ is the set of literals appearing within the scope of the `at end` temporal specifier in the action's `:condition` or `:duration` formulas.

- $\text{add}(a_{\text{end}})$, $\text{del}(a_{\text{end}})$ and $\text{NE}(a_{\text{end}})$ are the sets of added and deleted facts and numeric effects appearing within the scope of the `at end` temporal specifier in the action's `:effect`.

- $\text{pre}(a_{\text{overall}})$ is the set of literals appearing within the scope of the `over all` temporal specifier in the action's `:condition`.

The two actions a_{start} and a_{end} that represent the start and end events of a are sometimes referred to as "*snap actions*". The action a_{overall} has no effects; it's role in the simple plan is only to verify that each action's invariant conditions hold throughout its execution.

Note that the partitioning of a durative action's condition into $\text{pre}(a_{\text{start}})$, $\text{pre}(a_{\text{end}})$, and $\text{pre}(a_{\text{overall}})$ presupposes that the condition is a conjunction of these parts. PDDL, currently, does not provide any interpretation of a disjunctive condition over different points in time, such as, for example, `(or (at start (paid)) (at end (paid)))`. In fact, the article by Fox and Long [2003] restricts the condition of a durative action to be only a conjunction of literals, though it is straightforward to generalise the definitions to deal with arbitrary condition formulas at each single time point.

Definition 5.3   Let S be a temporal plan. The set of *happening time points* is

$$HT(S) = \{t \mid t : a[d] \in S\} \cup \{t + d \mid t : a[d] \in S\}.$$

Furthermore, let $t_1, \ldots, t_k$ be the time points in $HT(S)$ sorted in increasing order. The *induced simple plan*, $\text{SimplePlan}(S)$, is a multiset of timed instantaneous actions which includes:

- $t : a_{\text{start}}$ `[?duration=d]` and $t + d : a_{\text{end}}$ `[?duration=d]` for each timed action occurrence $t : a[d] \in S$, where a `[?duration=d]` denotes action a with the `?duration` parameter replaced with constant d; and

- $\left(\frac{t_i + t_{i+1}}{2}\right) : a_{\text{overall}}$ for each timed action occurrence $t : a[d] \in S$ and $t_i, t_{i+1} \in HT(S)$ such that $t \leq t_i$ and $t_{i+1} \leq t + d$.

Note how the induced simple plan "inserts" occurrences of a_{overall} in between every pair of happening time points that fall between the start and end of the action. This is because at every happening, the state may change, and the action's `over all` condition must hold in every state during the time that the action is ongoing.

Although the induced simple plan is still a collection of time stamped action occurrences, because these are no longer durative actions, the only remaining function of the time stamps is to order these action occurrences. That is, we can view the simple plan (almost) as a sequence of ground actions. The executability and validity of this sequence is analogous to that of a (classical or numeric) sequential plan: the sequence of action sets induces a corresponding sequence of states (cf. Definitions 2.8 and 4.5), and the plan is executable if each action's precondition holds in the state where it is applied and valid if in addition the goal condition holds in the final state (cf. Definitions 2.9 and 4.6). The only added complication here is that we can have several instantaneous actions occurring at the same point in time. For the induced state sequence to be unambiguous, it is required that all actions occurring at the same time are *non-interfering*. We already presented the classical definition of non-interference, Definition 2.10, when discussing partially ordered plans in Section 2.2.2. However, PDDL version 2.1 adopted a definition of non-interference that is more restrictive, and that also accounts for numeric conditions and effects:

Definition 5.4   Two ground instantaneous actions a_1 and a_2 are *non-interfering* iff (*i*) $(\text{pre}(a_1) \cup \text{add}(a_1)) \cap \text{del}(a_2) = (\text{pre}(a_2) \cup \text{add}(a_2)) \cap \text{del}(a_1) = \emptyset$; (*ii*) $\text{add}(a_1) \cap \text{pre}(a_2) = \text{add}(a_2) \cap \text{pre}(a_1) = \emptyset$; (*iii*) no ground function term appearing on the left-hand side in a numeric effect in $\text{NE}(a_1)$ appears anywhere in the precondition of a_2, and vice versa; and (*iv*) the combined numeric effects $\text{NE}(a_1) \cup \text{NE}(a_2)$ is non-conflicting according to Definition 4.4 (that is, for every ground function term f there is at most one effect on f in $\text{NE}(a_1) \cup \text{NE}(a_2)$ or all effects on f are additive).

Note that an action *is* permitted to interfere with itself in ways that are not permitted between actions occurring at the same time. For example, we have seen several examples of actions that negate some part of their own precondition at the start of the action; this is allowed. But a different action negating the same precondition may not occur at the same time. Likewise, an action may have an effect on a ground function term that appears in its precondition, but it can not be concurrent with another action that affects the same function term.

Definition 5.5   A temporal plan S is *valid* iff (*i*) for any pair of simultaneous actions $t : a_1$ and $t : a_2$ in SimplePlan(S), a_1 and a_2 are non-interfering; and (*ii*) any sequencing of the actions in SimplePlan(S) by increasing time stamps is a valid classical plan.

Same as in the case of classical planning, if a set of non-interfering instantaneous actions are all applicable in a state, they can be executed in any order with the same end result. Thus, the validity of a temporal plan does not depend on how actions in SimplePlan(S) occurring at the same time are ordered (if any sequencing is valid, all of them are).

There are a few complex issues in the semantics of temporal plans that we have not touched on above. We briefly outline them in the remainder of this section.

ϵ Separation

In the construction of the induced simple plan, any two actions that do not occur at the same time must be separated by a non-zero interval of time. The minimum time separating two nonsimultaneous actions is often referred to as ϵ ("epsilon"). PDDL version 2.1 does not specify what the value of ϵ is, only that it is an implementation-dependent constant greater than zero. Temporal planners that implement the PDDL version 2.1 semantics, and the VAL plan validator, require a concrete value. A mismatch in the value of ϵ can often be the reason why a seemingly correct temporal plan fails to validate.

The strong non-interference requirement (Definition 5.4) means that if action a adds some fact that is a precondition of another action a', they cannot be simultaneous. This means that if $t{:}a$ is the action that achieves a fact f, and f is a precondition of a', then a' cannot occur earlier than $t + \epsilon$ in the plan. For example, in the plan shown in Figure 5.3, the action (move-down e2 n3 n2) has the effect (at end (lift-at e2 n2)) and the action (move-down e2 n2 n1) has the condition (at start (lift-at e2 n3)). Thus, if the former action ends at time 5.667, and $\epsilon = 0.001$, the latter action can only start at 5.668, ending at 6.668. The action (leave p3 n1 e2), on the other hand, has the condition (over all (lift-at e2 n3)). Because this condition is not required to hold at the start of the action, only at a point in between each happening time point in the interior of its interval of execution, it can start immediately at time 6.668.

Some temporal planners (and the VAL plan validator!) actually implement a weaker notion of non-interference that removes condition (*ii*) from Definition 5.4. However, the ϵ separation between the effect that achieves a fact and an action whose precondition requires it to hold remains also in this (somewhat non-standard) interpretation. Thus, it allows the actions a and a' in the example above to be simultaneous, but only if the fact f required by pre(a') is already true beforehand.

Accounting for TILs

In the definitions above we have not included TILs, but this can be done easily by adding for each TIL (at t l) an instantaneous action t : a_f to SimplePlan(S), where a_f has no precondition and the effect of asserting the literal l. Note that in a temporal planning problem with TILs, the goal must still hold in the final state. This means that if the goal is achieved before some TILs occur, it must remain true after they have occurred.

The presence of TILs, however, complicate determining the total duration (makespan) of a plan, since TILs may still take place long after the last action in the plan has finished, and the goal has been achieved. Edelkamp and Hoffmann [2004], who introduced TILs, define the makespan of a temporal plan as the earliest time that is after the execution of all actions in the plan have finished and such that the goal condition is true from that time on. In other words, TILs that occur after the plan's actions and that do not contribute to achieving the goal are not considered part of the plan for the purpose of determining its makespan.

Durative Actions with Conditional Effects

Durative actions can have conditional effects, as described in Chapter 3. In a durative action, the condition and effect parts of a conditional effect are each given a temporal specifier. As long as both parts refer to the same time, as in, for example,

```
(when (at start (passenger-at ?pax ?floor))
      (at start (not (passenger-at ?pax ?floor))))
```

they can be straightforwardly assigned to either a_{start} or a_{end}, and handled the same as conditional effects in a non-temporal plan. The case when the condition and effect parts refer to different time points is more complex. Conditional effects where the effect takes place at a time before the time at which the condition is checked, as in

```
(when (at end (lift-at ?lift ?floor))
      (at start (boarded ?pax ?lift)))
```

are simply ruled out from the language [Fox and Long, 2003]. Conditional effects in which the condition is checked at an earlier time than the effect takes place, as in

```
(when (and (at start (passenger-at ?pax ?floor))
           (over all (lift-at ?lift ?floor)))
      (at end (boarded ?pax ?lift)))
```

are allowed. Such effects introduce a "memory" into the induced simple plan, linking the effect at one point in time to the state at an earlier point in time. However, because the planning model is deterministic the state that decides whether the conditional effect occurs or not is predictable; this means that the compilation described in Chapter 3 that replaces an action with a conditional effect with alternative actions representing the case when the effect fires and when it does not is still possible.

5.4   COMBINING NUMERIC AND TEMPORAL PLANNING

Combining the numeric planning features of PDDL discussed in Chapter 4 with temporal planning allows modelling of numeric change over time, which is particularly useful when resources are involved in the planning problem.

In the previous section, we already accounted for durative actions having discrete numeric effects, meaning effects that take place at the start or at the end of a durative action. However, PDDL also includes a mechanism to express *continuous* numeric effects of durative actions, meaning that the values of numeric state variables change continuously over the duration of the action. In this section, we introduce an example modelling a numeric-valued resource. First, we show a conservative approximation using discrete effects, and later a more accurate version using continuous effects.

Consider the example of driving a car between two locations, and suppose that we want to enrich the domain by taking fuel consumption into account. For the sake of simplicity let us assume that the amount of fuel consumed is proportional to the travel time. This can be modelled as follows:

```
(:durative-action drive
 :parameters (?c - car ?l1 ?l2 - loc)
 :duration (= ?duration (drive_time ?l1 ?l2))
 :condition (and (at start (car_at ?c ?l1))
                 (at start (>= (fuel ?c)
                              (* (drive_time ?l1 ?l2) (fuel_rate ?c))))
                 (over all (road_open ?l1 ?l2))
                 (at end (space_at_loc ?l2)))
 :effect (and (at start (decrease (fuel ?c)
                              (* (drive_time ?l1 ?l2) (fuel_rate ?c))))
              (at start (not (car_at ?c ?l1)))
              (at start (space_at_loc ?l1))
              (at end (car_at ?c ?l2))
              (at end (not (space_at_loc ?l2)))))
)
```

PDDL Example 15: Durative action for driving a car between two locations with fuel consumption.

There are two things to be noted in this example: first, taking resource consumption into account implies that there must be enough resources before an action can be applied. For this reason we added to the action condition that there must be enough fuel in the car at the start of the action. Similarly, we added to the action effects that the amount of fuel is decreased proportionally to the action duration.

The second thing to note is the notion of *conservative resource updating*. Indeed, the decreasing of the fuel is modelled as an `at start` effect, even though it happens continually over the duration of the action. The conservative modelling of resource updating requires that consumption of a resource is applied at the start of the durative action, while the production of a resource is applied at the end of the durative action.

Following the conservative model, we can now add an action for refueling the car as follows:

```
(:durative-action refuel
 :parameters (?c - car ?l1 - loc)
```

```
:duration (= ?duration 120)
:condition (and (at start (car_at ?c ?l1))
                (at start (petrol_pump_at ?l1)))
:effect (and (at end (assign (fuel ?c) (capacity ?c))))
)
```

PDDL Example 16: Durative action for refueling a car.

The importance of the conservative resource updating model is that it ensures that no parallel activities will consume resources that have already been committed to other activities. While this assumption is fine for a large number of problems, there are cases where the modelling of consumption and production of resources must be more accurate, and this requires the modelling of continuous change.

PDDL2.1 uses #t to refer to the continuously changing time from the start of a durative action during its execution. If we wanted to revise the drive action of Example 15 and take the continuous consumption of fuel into account, we would have the following action:

```
(:durative-action drive
 :parameters (?c - car ?l1 ?l2 - loc)
 :duration (= ?duration (drive_time ?l1 ?l2))
 :condition (and (at start (car_at ?c ?l1))
                 (at start (>= (fuel ?c)
                               (* (drive_time ?l1 ?l2) (fuel_rate ?c))))
                 (over all (road_open ?l1 ?l2))
                 (at end (space_at_loc ?l2)))
 :effect (and (decrease (fuel ?c) (* #t (fuel_rate ?c)))
              (at start (not (car_at ?c ?l1)))
              (at start (space_at_loc ?l1))
              (at end (car_at ?c ?l2))
              (at end (not (space_at_loc ?l2)))))
)
```

PDDL Example 17: Durative action for driving a car between two locations with *continuous* fuel consumption.

Note that we do not specify the fuel consumption as either an at start or at end effect, because it is a continuous effect that is not temporally annotated. A mathematical interpretation of #t is the differential equation, as indeed the continuous consumption of fuel could be interpreted as

$$\frac{d}{dt}(\text{fuel ?c}) = (\text{fuel\_rate ?c})$$

Combining continuous change with duration inequalities adds significant expressive power. For example we can revise the refuel action in Example 16, where we made the simplistic assumption of having a fixed duration for the durative action whose effect is to refill the car at its full capacity. A more realistic scenario is where the refuel action can have a flexible duration, and the amount of refueling will depend on the duration, *which is chosen by the planner*.

```
(:durative-action refuel
 :parameters (?c - car ?l1 - loc)
 :duration (>= ?duration 0)
 :condition (and (at start (car_at ?c ?l1))
                 (at start (petrol_pump_at ?l1))
                 (overall (<= (fuel ?c) (capacity ?c)))
                 )
 :effect (and (increase (fuel ?c) (* #t (refuel_rate ?c))))
 )
```

PDDL Example 18: Durative action for refueling a car with duration inequality.

Note that with this model the planner has control over how long to recharge the car, and this is particularly interesting in problems where the objective is to minimize the makespan and therefore the planner can plan to refuel the car with the minimum amount of fuel needed to finally achieve the goal. Note also that the over all condition prevents the planner for exceeding the maximum capacity.

All the effects describing continuous change must take the form of (* #t <*quantity*>) where <*quantity*> can be any numeric expression, and hence this also allows the modelling of complex non-linear dynamics.

Furthermore, it is possible to describe even more complex forms of continuous change by using interdependent concurrent effects, that may lead to continuous change described by logarithmic and exponential functions.

A simple example of interdependent concurrent effects is given by the interaction between distance traveled, velocity and acceleration of a car that could be modelled as

```
(and (increase (distance ?c) (* #t (velocity ?c)))
     (increase (velocity ?c) (* #t (acceleration ?c))))
```

For the full semantics of temporal planning with continuous effects, we refer the reader to the article by Fox and Long [2003].

CHAPTER 6

Planning with Hybrid Systems

Hybrid systems are systems described by discrete as well as continuous variables. Indeed, many real-world applications present such a hybrid dynamics. Moreover, many of our interactions with the world involve us performing actions that initiate, terminate, control, or simply avoid continuous processes, and those scenarios are very common particularly in robotics or embedded systems. Processes can be seen as *flows* (of heat, energy, liquids, gases, traffic, money, information, etc.).

In hybrid systems, the executive agents perform actions that initiate or terminate processes. Planning with hybrid systems, therefore, requires the planner to anticipate how the effects of the actions will interact with the processes.

PDDL+ is the extension of PDDL for modelling hybrid systems through the use of *continuous processes* and *events*, and was designed by Maria Fox and Derek Long. A full description of the language with its semantics can be found in Fox and Long [2006].

In order to help the reader to familiarize with this complex language, we begin by presenting some examples used to introduce the main components of PDDL+, namely continuous processes in Section 6.1, exogenous events in Section 6.2, and their combinations in Sections 6.3 and 6.4. We conclude the chapter with a brief discussion in Section 6.5 of plan validity in hybrid domains. Note that most of the material presented in this chapter is based on papers and talks by (and conversations with) Maria Fox and Derek Long.

6.1 CONTINUOUS PROCESSES

We begin with an example designed by Derek Long and presented in the tutorial on Planning in Hybrid Domains at ICAPS 2013. Let us consider the problem of modelling the effect of a robot hand dropping a ball [Fox et al., 2005]. As discussed in the previous chapter, continuous change within the durative actions can be modelled as follows:

```
(:durative-action drop-ball
 :parameters (?b - ball)
 :duration (> ?duration 0)
 :condition (and (at start (holding ?b)) (at start (= (velocity ?b) 0))
                 (over all (>= (height ?b) 0)))
 :effect (and (at start (not (holding ?b)))
```

```
              (decrease (height ?b) (* #t (velocity ?b)))
              (increase (velocity ?b) (* #t (gravity)))))
```

PDDL Example 19: A durative action to model a ball dropping.

Here the action duration is not fixed, and the preconditions state that in order to apply this action the robot hand must be holding the ball, and that the ball is initially stationary. Immediately after we apply the action, the robot is no longer holding the ball, and two continuous changes affect the ball. Namely, as the ball is falling, the height of the ball is decreasing depending on velocity ($dh/dt = v$) and velocity is increasing depending on gravity ($dv/dt = g$).

Note that the additional invariant condition (over all (>= (height ?b) 0)) is used to stop the effect of the action when the ball reaches the ground. However, this action does not allow an easy modelling of what happens if the ball is caught before reaching the ground, or what happens if the ball bounces.

A better model is to consider the action of releasing the ball as separated from what happens to the ball when it falls. In this sense, a *discrete* action (drop the ball) initiates a *continuous* change (the ball falling), that can be terminated by various possible actions (reaching the floor, being caught, bouncing, etc.). PDDL+ allows this separation through the use of continuous processes.

A process in PDDL+ becomes active as soon as the preconditions become true, and remains active as long as the preconditions are not violated. Processes are not directly applied by the planner, and they do not appear explicitly in the plan. However, processes are triggered as a consequence of actions in the plan, and this makes PDDL+ planning hard as the planner has to anticipate the activation and deactivation of processes as a consequence of actions in the plan being executed. Also, note that multiple processes can be active at the same time. Examples are provided later in this chapter.

Going back to the example of the robot hand dropping the ball, in the following we show how this example can be modelled in PDDL+. First, the discrete action for releasing the ball is as follows.

```
(:action release
 :parameters (?b - ball)
 :precondition (and (holding ?b) (= (velocity ?b) 0))
 :effect (and (not (holding ?b))))
```

PDDL Example 20: A discrete action to model the release of the ball.

The only effect of this action is to delete (holding ?b), and this, in turn, will trigger the process of the ball falling, which is modelled as follows:

```
(:process fall
 :parameters (?b - ball)
 :precondition (and (not (holding ?b)) (> (height ?b) 0))
 :effect (and (increase (velocity ?b) (* #t (gravity)))
              (decrease (height ?b) (* #t (velocity ?b)))))
```

PDDL Example 21: A continuous process to model the ball falling.

Assuming that the ball is not on the ground (height > 0), immediately after the action release is applied, the process preconditions become true and the process is triggered. The effects of the process model the continuous change, same as in the durative action in Example 19. The termination of the process is no longer linked to the action that triggered it, but only to the process preconditions. Thus it is possible, for example, to extend the domain with the following action to catch the ball (and stop the falling process).

```
(:action catch-ball
 :parameters (?b - ball)
 :precondition (and (>= (height ?b) 5) (<= (height ?b) 5.01))
 :effect (and (holding ?b) (assign (velocity ?b) 0)) (caught ?b))
```

PDDL Example 22: An action to catch the ball.

This action models the situation in which the robot hand is positioned 5 meters above the ground, and can only grasp the ball when it is within a narrow height range. The robot (i.e., the planner) has to find the exact time when the ball is at the same height so that the hand can catch it.

The last step is to model the bouncing of the ball when it reaches the ground, and this can be done using PDDL+ *events*.

6.2   EXOGENOUS EVENTS

Events in PDDL+ are used to model changes in the environment, and in this sense they are considered "the equivalent of actions for the world" [Fox et al., 2005]. They are triggered as soon as their preconditions become true, and their effects are instantaneous. As for processes, events are not directly applied by the planner, and they do not appear explicitly in the plan.

An event is what we need to model the ball bouncing, as this is something that happens because of interactions in the environment. The event can be written as follows:

```
(:event bounce
 :parameters (?b - ball)
 :precondition (and  (>= (velocity ?b) 0) (<= (height ?b) 0))
 :effect (and (assign (height ?b) (* -1 (height ?b)))
              (assign (velocity ?b) (* -1 (velocity ?b)))))
```

PDDL Example 23: An event to model the ball bouncing.

The ball will bounce as soon as it touches the ground (note that the precondition requires `height` ≤ 0 rather than `height` $= 0$ to handle numerical imprecision). The effect of triggering the bounce event is changing the current sign of velocity and height (that again could be less than zero due to numerical imprecision).

Let us assume we have the following problem instance for the ball domain:

```
(define (problem ball-problem)
    (:domain ball)

    (:objects b1 - ball)

    (:init (= (height ?b1) 10)
           (= (velocity ?b1) 0)
           (= (gravity) 9.8)
           (holding ?b1))

    (:goal (and (caught ?b1)))
```

Then a valid plan is the following:

```
0.100: (release b1)
4.757: (catch-ball b1)
```

When dealing with hybrid systems, and because of the interactions between discrete and continuous change, it is not always immediate to state whether a plan is valid or not. This is one example, where it is not even trivial to say how many times the ball bounces before being caught. This is why being able to validate plans is key [Howey et al., 2004] (a validation for this plan is shown in Section 6.5).

6.3   EXAMPLE: THE GENERATOR

The generator is one of the first benchmarks designed for PDDL+, proposed by Maria Fox, Richard Howey, and Derek Long. The problem is the following: *a generator must run for X seconds. It is powered by a fuel tank with a limited capacity of Y fuel units and consumes one fuel unit per second. Z fuel tanks of K fuel units can be used to refuel the generator at a rate of 2 fuel units per second.* Several variants of this problem have been proposed [see for example Fox et al. [2003]] with different types of dynamics, including linear and nonlinear continuous change. The challenge here is given by the concurrency involved in the model, as refuelling tanks can be used in parallel, while also the generator is running. Hence, the generator is a good example of how to model *concurrent processes*.

The domain is as follows:

```
(define (domain generator)
  (:requirements :fluents :durative-actions
   :adl :typing :time)

  (:types generator tank)

  (:predicates (generator-ran) (available ?t - tank)
               (using ?t - tank ?g - generator)
               (safe ?g - generator))

  (:functions (fuelLevel ?g - generator) (capacity ?g - generator)
              (fuelInTank ?t - tank) (target ?g - generator)

  (:durative-action generate
   :parameters (?g - generator)
   :duration (= ?duration (target ?g))
   :condition (and (over all (>= (fuelLevel ?g) 0)) (over all (safe ?g)))
   :effect (and (decrease (fuelLevel ?g) (* #t 1))
               (at end (generator-ran)))
  )

  (:action refuel
   :parameters (?g - generator ?t - tank)
   :precondition (and (not (using ?t ?g)) (available ?t))
   :effect (and (using ?t ?g) (not (available ?t)))
  )

  (:process refuelling
```

```
    :parameters (?g - generator ?t -tank)
    :precondition (and (using ?t ?g))
    :effect (and (increase (fuelLevel ?g) (* #t 2))
                 (decrease (fuelInTank ?t) (* #t 2)))
  )

  (:event tankEmpty
    :parameters (?g - generator ?t - tank)
    :precondition (and (using ?t ?g) (<= (fuelInTank ?t) 0))
    :effect (and (not (using ?t ?g)))
  )

  (:event generatorOverflow
    :parameters (?g - generator)
    :precondition (and (> (fuelLevel ?g) (capacity ?g)) (safe ?g))
    :effect (and (not (safe ?g)))
  )
)
```

Note that generate is a durative action where the duration parameter will be instantiated with the goal (*running for X seconds*). The instantaneous action refuel will trigger the process refuelling that will be terminated by the event tankEmpty.

An example problem is as follows, where $X = 100$, $Y = 60$, $Z = 2$, $K_1 = 25$, and $K_2 = 15$.

```
(define (problem run-generator)
    (:domain generator)
    (:objects gen - generator tank1 tank2 - tank)
    (:init (= (target gen) 100)
           (= (fuelLevel gen)  60)
           (= (capacity gen)   60)
           (= (fuelInTank tank1) 25)
           (= (fuelInTank tank2) 15)
           (available tank1)
           (available tank2)
           (safe gen)
    )
    (:goal (generator-ran))
  )
```

A valid plan for this problem is:

```
000: (generate generator) [100]
059: (refuel generator tank1)
075: (refuel generator tank2)
```

Figure 6.1 shows the graphical view generated by VAL of the execution of this plan. This illustrates the behaviour of the interacting processes.

6.4   EXAMPLE: MULTIPLE-BATTERY MANAGEMENT

The following examples is taken from Fox et al. [2012] and is about the efficient use of multiple batteries, that is a practical problem with wide and growing application. When a system is powered by multiple batteries, the load is switched between batteries using a control system. An important issue is to devise efficient an switching strategy, as this can result in extending the total lifetime of the batteries quite significantly. In Fox et al. [2012], the problem of finding a good strategy has been addressed using planning, and PDDL+ is used to describe the complex dynamics of batteries. The following description of the example is taken from Fox et al. [2012].

6.4.1   KINETIC BATTERY MODEL

In the Kinetic Battery Model [Jongerden and Haverkort, 2009, Manwell and McGowan, 1993] the battery charge is distributed over two wells: the available-charge well and the bound-charge well.

A fraction c of the total charge is stored in the available-charge well, and a fraction $1 - c$ in the bound-charge well. The available-charge well supplies electrons directly to the load ($i(t)$, where t denotes the time), whereas the bound-charge well supplies electrons only to the available-charge well. The charge flows from the bound-charge well to the available-charge well through a "valve" with fixed conductance, k. Moreover, the rate at which charge flows between the wells depends on the height difference between the two wells. The heights of the two wells are given by:

$$h_1 = \frac{y_1}{c} \qquad h_2 = \frac{y_2}{1-c},$$

where y_1 is the the available charge, y_2 is the bound charge, and c is a constant depending on the characteristics of the battery. When a load is applied to the battery, the available charge reduces, and the height difference between the two wells grows. When the load is removed, charge flows from the bound-charge well to the available-charge well until the heights are equal again. The change in the charge in both wells is given by the following system of differential equations:

$$\begin{cases} \frac{dy_1}{dt} = -i(t) + k(h_2 - h_1) \\ \frac{dy_2}{dt} = -k(h_2 - h_1) \end{cases}$$

with initial conditions $y_1(0) = c \cdot C$ and $y_2(0) = (1 - c) \cdot C$, where C is the total battery capacity.

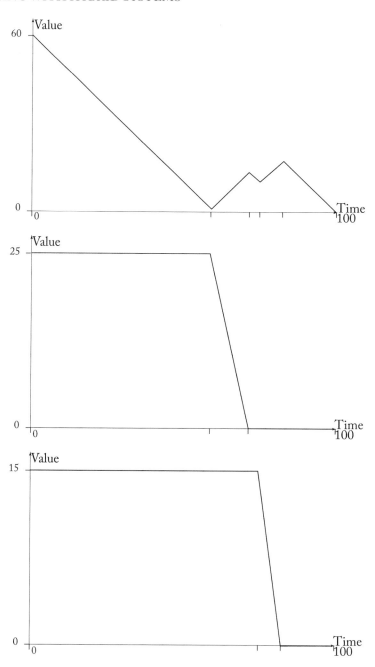

Figure 6.1: **VAL** report for Generator. Top graph corresponds to the generator fuel level. The graph below represents tank1 fuel level, and at the bottom the fuel level of tank2.

To describe the discharge process of the battery, as in Jongerden et al. [2009], we adopt coordinates representing the height difference between the two wells, $\delta = h_2 - h_1$, and the total charge in the battery, $\gamma = y_1 + y_2$. In this new setting $y_1 = c(\gamma - (1-c)\delta)$.

The change in both wells is then given by the system of differential equations

$$\begin{cases} \frac{d\delta}{dt} = \frac{i(t)}{c} - k'\delta \\ \frac{d\gamma}{dt} = -i(t) \end{cases}$$

with solutions

$$\begin{cases} \delta(t) = \frac{i}{c} \cdot \frac{1-e^{-k't}}{k'} \\ \gamma(t) = C - it \end{cases} ,$$

where $k' = k/(1-c)c$, $\delta(0) = 0$, and $\gamma(0) = C$. The condition for a battery to be empty is $\gamma(t) = (1-c)\delta(t)$, at which point the battery is considered to be dead as it can't be used and can't recover charge anymore.

6.4.2 PDDL+ MODEL FOR THE KINETIC BATTERY

The dynamics of Kinetic Battery can be captured very easily in PDDL+. The domain contains:

- the two processes `consume` and `recover` that describe the dynamics of the batteries;

- a durative action `use` of variable duration that allows the planner to select which battery to use over an interval; and

- an event `batteryDead` that is triggered once the available charge of a battery is exhausted.

  This is an interesting example of how to model interactions between actions, processes, and events.

```
(define (domain battery)
  (:requirements :fluents :durative-actions :duration-inequalities
   :adl :typing :time :timed-initial-literals)

  (:types battery)

  (:predicates (switchedOn ?b - battery) (switchedOff ?b - battery)
               (dead ?b - battery) (satisfactoryService)
               (allDone) (finished))

  (:functions (delta ?b - battery) (gamma ?b - battery) (load)
              (cParam ?b - battery) (kprime ?b - battery) (services))
```

```
(:process consume
 :parameters (?b - battery)
 :precondition (switchedOn ?b)
 :effect (and (decrease (gamma ?b) (* #t (load)))
              (increase (delta ?b) (* #t (/ (load) (cParam ?b)))))
)

(:process recover
 :parameters (?b - battery)
 :precondition (>= (delta ?b) 0)
 :effect (and (decrease (delta ?b) (* #t (* (kprime ?b) (delta ?b)))))
)

(:durative-action use
 :parameters (?b - battery)
 :duration (>= ?duration 0)
 :condition (and (at start (switchedOff ?b))
                 (over all (switchedOn ?b))
                 (over all (not (dead ?b))))
 :effect (and (at start (and (switchedOn ?b) (not (switchedOff ?b))
              (increase (services) 1)))
              (at end (and (switchedOff ?b) (not (switchedOn ?b))
              (decrease (services) 1))))
)

(:event batteryDead
 :parameters (?b - battery)
 :precondition (and (switchedOn ?b)
                    (<= (gamma ?b) (* (-1 (cParam ?b)) (delta ?b))))
 :effect (and (not (switchedOn ?b)) (dead ?b))
)

(:action completeService
 :parameters (?b - battery)
 :precondition (and (switchedOn ?b) (finished))
 :effect (and (allDone) (switchedOff ?b) (not (switchedOn ?b)))
```

```
  )
)
```

PDDL Example 24: The *Multiple-Battery* domain.

The load profile to be serviced is encoded in the PDDL+ problem through the use of *timed initial literals*, which allow expression of exogenous events corresponding, in this case, to changes in the load value. This is an another example of how TILs can be used, i.e., to specify the input of the planning problem, in addition to the cases discussed in Chapter 5. A fragment of the problem (which also contains the battery specification) is shown below:

```
(define (problem 2B)
  (:domain kibam)

  (:objects b1 b2 - battery)

  (:init
    (= (cParam b1) 0.166)
    (= (kprime b1) 0.122)
    (= (gamma b1) 5.5)
    (= (delta b1) 0)

    (at 0 (= (load) 0.25))
    (at 1.00 (= (load) 0.50))
    (at 2.00 (= (load) 0.25))
    (at 3.00 (= (load) 0.50))
    (at 4.00 (= (load) 0.25))
    ...
    (at 90.0 (= (load) 0.50))
    (at 90.1 (finished))
  )
  (:goal (allDone))
)
```

PDDL Example 25: Fragment of the problem for the *Multiple-Battery* domain.

Note that for setting the goal, a time literal is used in the initial state to add (finished) at the end of the load which allows the planner to apply the action completeService to achieve the goal.

6.5   PLAN VALIDATION IN HYBRID DOMAINS

Like in temporal planning (cf. Chapter 5), a plan for a hybrid planning problem is a set of time-stamped action occurrences, which induces a series of happenings. The presence of exogenous events and processes in PDDL+, however, means that additional happenings, where events fire or processes start or stop, may appear in the execution of the plan. In between happenings, the state evolves continuously according to the set of active processes and the continuous effects of on-going actions.

For the general set of continuous dynamics that can be written in PDDL+, the problem of deciding if a numeric condition holds throught an interval of continuous change—known as the "zero-crossing problem" in the literature—is undecidable. Since solving this problem this is necessary to determine where to insert event firings in a plan's execution, validation of plans for PDDL+ domains is undecidable in general. Analytical solutions exist for restricted subsets of continuous effects [Howey et al., 2004], and approximate numerical methods can be used in the general case, although such methods can not guarantee the plan's validity. Events raise further difficulties, for example cascades of events triggering without separation in time, or an infinite series of events occurring in a bounded interval of time (so-called "zenoness") [Fox et al., 2005].

The plan validator in the VAL tool suite, which we mentioned already in Chapter 2, implements plan validation for a subset of PDDL+ domains. It can provide a trace showing which processes and events are triggered while the plan executes. A fragment of the validator report for the bouncing ball problem (from Section 6.2) is shown in Figure 6.2. Note that it is only the happening at time 4.757 that corresponds to an action in the plan. It can also output a graph of the continuous evolution of numeric state variables during the plan's execution. As an example, Figure 6.3 shows the evolution of (height b1), showing that the ball bounced twice before being caught.

At this point, one may note that the current model of the bouncing ball assumes a perfectly elastic collision between the ball and the floor. One way to model a non-perfectly elastic collision is as follows:

```
(:event bounce-non-perf-elastic
 :parameters (?b - ball)
 :precondition (and   (>= (velocity ?b) 0)
                      (<= (height ?b) 0.00001))
 :effect (and (assign (height ?b) (* -1 (height ?b)))
              (assign (velocity ?b) (* (coeffRest ?b) (velocity ?b)))))
```

1.51421: Event triggered!
Triggered event (bounce b1)
Unactivated process (fall b1)
Updating **(height b1)** (-2.22045e-15) by 2.22045e-15 assignment.
Updating **(velocity b1)** (14.1421) by -14.1421 assignment.

1.51421: Event triggered!
Activated process (fall b1)

4.34264: Checking Happening... ...OK!
(height b1)$(t) = -5t^2 + 14.1421t + 2.22045e - 15$
(velocity b1)$(t) = 10t - 14.1421$
Updating **(height b1)** (2.22045e-15) by -2.44943e-15 for continuous update.
Updating **(velocity b1)** (-14.1421) by 14.1421 for continuous update.

4.34264: Event triggered!
Triggered event (bounce b1)
Unactivated process (fall b1)
Updating **(height b1)** (-2.44943e-15) by 2.44943e-15 assignment.
Updating **(velocity b1)** (14.1421) by -14.1421 assignment.

4.34264: Event triggered!
Activated process (fall b1)

4.757: Checking Happening... ...OK!
(height b1)$(t) = -5t^2 + 14.1421t + 2.44943e - 15$
(velocity b1)$(t) = 10t - 14.1421$

Updating **(height b1)** (2.44943e-15) by 5.00146 for continuous update.
Updating **(velocity b1)** (-14.1421) by -9.99854 for continuous update.

4.757: Checking Happening... ...OK!
Adding (holding b1)
Updating **(velocity b1)** (-9.99854) by 0 assignment.

4.757: Event triggered!
Unactivated process (fall b1)

Plan executed successfully - checking goal

Figure 6.2: Fragment of triggered events and active processes provided by VAL.

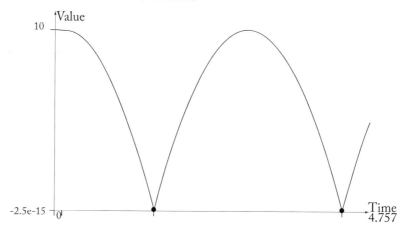

Figure 6.3: Graphical representation of continuous change in the variable height (y-axis represents meters) provided by VAL.

A coefficient of restitution is added that will decrease the velocity of the ball at each bounce. Furthermore, we also relax the precondition on height to allow bouncing to happen until the ball is too close to the floor to bounce any more.

The evolution of the value of (height b1) when the plan is executed in this domain is shown in Figure 6.4. (Note that with the new domain, the previous plan is no longer valid and the robot does not catch the ball!)

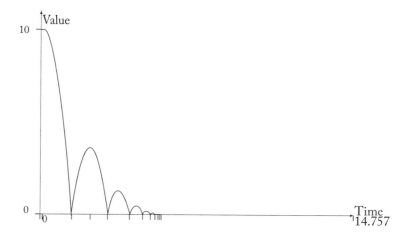

Figure 6.4: VAL report for non-perfectly elastic collision. The graph shows the change of variable height over time (y-axis represents meters).

6.5.1   MODELLING ASSUMPTION IN PDDL+

We conclude this section by briefly recalling some assumptions about the class of domains that can be modelled in PDDL+, which are described in more detail by Fox and Long [2006].

- **No Moving Targets** and **Epsilon Separation**: The no moving target rule states that no two actions are allowed to simultaneously make use of a value if one of the two is accessing the value to update it (i.e., the value is a moving target for the other action to access). As a consequence of this restriction, plans must respect the *epsilon-separation* requirement, i.e., interfering actions must be separated by at least a time interval of length ϵ.

- **Events**: The PDDL+ events are particularly challenging as they could trigger an infinite cascading sequence of events. To address this issue, the following assumptions are made in PDDL+. First, each event must delete one of its own preconditions and thus avoid self triggering. Second, planning instances must be *event-deterministic*: in every state in which two events e_1 and e_2 are applicable, the transition sequences e_1 followed by e_2 and e_2 followed by e_1 are both valid and reach the same resulting state. Note that the condition for ensuring this is similar to what is described by items (i), (iii), and (iv) of 5.4 for non-interfering actions.

CHAPTER 7

Conclusion

The Planning Domain Definition Language (PDDL) has been developed and increasingly adopted by the AI planning community for over two decades. Throughout this text, we have seen a range of the more commonly used subsets of PDDL, as well as a variety of modelling examples demonstrating the expressiveness of each. The widespread use of PDDL stems from the collaborative efforts of the planning community, and the desire for a language to facilitate benchmarking and application of planning systems. We have highlighted some of the issues and difficulties with the language throughout this text, but the usefulness of having a common language for much of the field so far outweighs the difficulties that arise from ambiguities in its specification.

There are several other extensions to the language that we did not cover, and we briefly detail some of them in Section 7.1. We wrap up this book in Section 7.2 with a discussion on some possible future paths for PDDL.

7.1 OTHER PLANNING PDDL-LIKE LANGUAGES

Since the introduction of PDDL for classical planning problems, various sub-fields of the planning community have invented variants of the language for expressing other types of planning problems. Typically, these extend the syntax of PDDL in a minor way to capture a particular phenomenon of planning. We will not provide a full exposition of the various settings, but instead give a flavour of how the research community has extended PDDL to accommodate some key additional features.

Fully Observable Non-Deterministic Planning

The setting of Fully Observable Non-Deterministic (FOND) Planning extends the classical planning formalism by allowing actions to have more than one outcome, potentially leading to more than one successor state when executed. Following the style of PDDL languages in general, the specification of the actions effects can be done in a factored manner. Introduced originally for International Planning Competition in 2008 [Bryce and Buffet, 2008], the standard PDDL language was extended to include the requirement `:non-deterministic`, and the effects could subsequently use the keyword `oneof` to indicate multiple possible effects that could occur. Crucially, the `oneof` clauses were permitted to be nested and combined. Consider the following example of a nondeterministic action:

```
(:action foo
  :parameters (...)
  :precondition (...)
  :effect (and (f1) (oneof (f2) (f3))
               (oneof (and) (f4
                      (oneof (f5) (f6)))))
```

PDDL Example 26: An example action schema for the FOND setting.

The nesting of `oneof` clauses will combine to explicitly represent a number of possible outcomes. Note, also, the use of an empty `and` clause to indicate that no part is selected for that part of the action effect. Written out explicitly, the above PDDL snippet is equivalent to the following:

```
(:action foo
  :parameters (...)
  :precondition (...)
  :effect (oneof (and (f1) (f2))
                 (and (f1) (f3))
                 (and (f1) (f2) (f4))
                 (and (f1) (f3) (f4))
                 (and (f1) (f2) (f5))
                 (and (f1) (f3) (f5))
                 (and (f1) (f2) (f6))
                 (and (f1) (f3) (f6))))
```

PDDL Example 27: An example action schema for the FOND setting without nesting.

Whereas most of the planning formalisms discussed in this text have a solution in the form of a sequence of actions (potentially timed), the presence of nondeterminism requires a plan to instead take the form of a decision tree, controller, or policy that maps the states of the world to the action that should be taken. This way, the various outcomes of a nondeterministic action can be considered independently.

Probabilistic PDDL

Like the nondeterministic extension of PDDL, the probabilistic extension (PPDDL) introduced by for the International Planning Competition in 2004 Younes and Littman [2004] allows for more than one state to be the result of applying an action, and so solutions take the form of controllers or policies. The keyword :probabilistic-effects is used to indicate the presence of probabilistic effects, and the syntax mirrors that of the oneof clauses with the addition of a probability term for each of the arguments:

(probabilistic <prob1> <effect1> <prob2> <effect2> ...)

The probability term can be either a floating point number (e.g., 0.25) or a fraction (e.g., 1/4), and just like oneof clauses, nesting is permitted. The following is a reduced example from an existing PPDDL benchmark that involves navigating between destinations with the possibility of navigation errors:

```
(:action drive-truck
  :parameters (?t - truck ?src - city ?dst - city)
  :effect (when (and (truck-at-city ?t ?src)
                     (can-drive ?src ?dst))
               (and (not (truck-at-city ?t ?src))
                    (probabilistic
                      0.8 (truck-at-city ?t ?dst)
                      0.2 (forall (?wrongdst1 - city)
                             (when (wrong-drive1 ?src ?wrongdst1)
                               (forall (?wrongdst2 - city)
                                 (when (wrong-drive2 ?src ?wrongdst2)
                                   (probabilistic
                                     1/2 (truck-at-city ?t ?wrongdst1)
                                     1/2 (truck-at-city ?t ?wrongdst2)
                                   )))))))))
```

PDDL Example 28: An example demonstrating probabilistic effects.

Note the mix of probability specification and nesting that is used. In total, there are three different outcomes with 80%, 10%, and 10% probability of occurring, respectively.

The problem file will specify the objective for the PPDDL problem, which is most commonly set to the following (along with the :rewards requirement):

```
(:goal-reward 1)
(:metric maximize (reward))
```

As noted at the beginning of this book, the most recent International Planning Competitions for probabilistic planning have shifted to using the RDDL language [Sanner, 2011].

Planning with Partial Observability and Sensing

For settings where the state of the world is only partially observable, fragments of PDDL were introduced to handle both uncertainty in the initial state and the result of sensing actions [Bonet and Geffner, 2001, Hoffmann and Brafman, 2006]. This formalism is commonly referred to as Planning with Partial Observability and Sensing (PPOS). Researchers have independently introduced a variety of approaches to model the uncertainty in the initial state, and we cover each of them briefly here.

For the specification of the initial state in the problem description, common patterns of specifying uncertainty in what is true or false are used. The first is an *exclusive or* on a set of predicates. As an example this stipulates that the location of a monster is known to be one of three possibilities:

```
(oneof (loc monster loc-1) (loc monster loc-4) (loc monster loc-13))
```

Similarly, an or clause in the initial states provides a traditional interpretation of having at least one of the arguments hold true. In both the `oneof` and or clauses, predicates may be negated as well. Despite the similarity to effect specifications in a nondeterministic setting, the agent will not know which predicate actually holds at planning or execution time. Instead, the truth of these predicates will be inferred through other means (sensing or deduction).

Because solvers of the PPOS formalism need to reason about the possible states the agent could be in, the initial state specification was also extended to allow for *invariants* to be specified: clauses that will hold true of every reachable state. These can be seen as explicit statements about the reachable state space, similar to the *state invariants* that we describe in Chapter 2. It is assumed that these invariants are correct and hold in every reachable state (as opposed to temporally extended goals or constraints that every plan must satisfy, as discussed in Chapter 3). Invariants allow the planner to infer extra information throughout the course of a plan. The syntax for this is very similar to the `oneof` and or clauses:

```
(invariant (loc monster loc-1) (loc monster loc-4) (loc monster loc-13))
```

Identifying a set of predicates as an invariant is stronger than using a `oneof` clause, as it holds in every reachable state and not just the initial state.

Finally, because the initial state in a PPOS problem represents uncertainty in what holds, predicates that are not explicitly mentioned are assumed to be unknown. This is in contrast with classical planning, where any predicate that is not mentioned is presumed to be false.

The other distinguishing aspect of the PPOS setting is observations that provide information about the state of the world at execution time. As an example, consider the following action from a domain where the location of a monster is unknown:

```
(:action smell_monster
 :parameters (?pos - pos)
 :precondition (and (alive) (at ?pos))
```

```
:observe (stench ?pos)
)
```

Note that instead of an :effect component to the action, we have :observe. This component of the action specification should indicate a single predicate that is known at execution time. In this example, when the smell_monster action is executed, the agent will know if (stench ?pos) is true or false. Another commonly used format for the sensing observation is to explicitly identify the action as a sensing one. The following example is equivalent to the one above, and used by a different variety of PPOS planners:

```
(:sensor smell_monster
 :parameters (?pos - pos)
 :condition (and (alive) (at ?pos))
 :sense (stench ?pos)
)
```

When there are no sensing actions as part of the domain model, solutions take the form of a classical plan (i.e., sequence of actions), and the fragment of planning is referred to as *conformant planning*. If, on the other hand, the agent is able to sense, solutions are similar to FOND—trees of execution, policies, or controllers are the typical representations used. This fragment of planning is typically referred to as either PPOS or *Contingent Planning*.

Multi-Agent Planning

A variety of planning approaches have been introduced to address the multi-agent setting, and many come with a custom syntax for representing this flavour of problem description. Two proposals in particular build on the PDDL syntax to define a multi-agent variant of planning [Brenner, 2003, Kovacs, 2012].

The first, by Brenner [2003], is an extension of PDDL 2.1 (cf. Chapter 4) that includes the following key features:

1. The expression of beliefs through the use of 3-valued logic (true, false, unknown), and multi-valued state variables.

2. A model of time that allows for both quantitative specification (as we have discussed in Chapter 5) and qualitative specification (as discussed in Section 3.4).

3. Allowing degrees of control over events and actions of other agents.

4. Plan synchronisation between agents through the notion of speech acts that may occur to convey truthful statements about the state of the world.

The following example from Brenner [2003] demonstrates some of these features:

```
(:durative-action Move_F_Loc2[Loc1_R12]
 :parameters (?a - agent ?dst - place)
 :duration (:= ?duration (interval 2 4))
 :condition (and (at start (== (pos F) Loc1))
                 (at start (== (connection Loc1 Loc1) R12))
                 (at start (clear R12)))
 :effect (and (at start (:= (pos ?a) R12))
              (at end (:= (pos ?a) Loc2)))
)
```

Note the special syntax for the `:duration` construct, as well as the assignment/equality syntax similar to PDDL's object fluents.

The second proposal, due to Kovacs [2012], is an extension of PDDL 3.1 (partially described in Chapter 2). The primary aim was to bridge the gap between the well-adopted single agent syntax of PDDL and the practice of implemented multi-agent systems. A driving principle behind the extension was to allow for planning both for other agents, as well as by the agents themselves (i.e., unifying the notions of centralised and distributed planning under a single umbrella). The following is a PDDL example from the original work:

```
(:action lift_table
 :agent ?a - agent
 :precondition (and (not (lifted table))
                    (at ?a table)
                    (exists (?b - agent)
                            (and (not (= ?a ?b))
                                 (at ?b table)
                                 (lift_table ?b))))
 :effect (lifted table)
 )
```

Note the dedicated specification of an `:agent` in the action schema, as well as the ability to quantify over other agents and describe their current action status.

7.2   THE FUTURE OF PDDL

Since its first appearance 20 years ago, PDDL has undergone several revisions. New language features have been added to allow it to express new kinds of planning problems, or to improve modelling convenience. Some of these extensions are now considered part of PDDL, while some are viewed as different, but related languages, for example the formalisms for expressing probabilistic and nondeterministic planning problems described earlier in this chapter. At the same time, some of the features initially proposed, such as the ability to define hierarchical task decompositions for example, were never used and have fallen out of the language.

Since PDDL is a community project, not a standard, and without any formal governance, we can only speculate on what may happen to it. It will depend on the interests of the AI planning research community, what kinds of planners are developed, and what their input language will look like.

INCREASING EXPRESSIVITY

There are many ways in which the expressivity of PDDL could grow. Some of these would lead to new classes of planning problems for the community to address.

While the classical fragment of PDDL has been relatively stable, some extensions have been proposed. Inspired by the *Functional* STRIPS formalism [Geffner, 2000], Helmert et al. [2008] introduced object-valued functions in PDDL 3.1, which allow modelling state variables with an arbitrary finite domain of discrete values instead of just the Boolean values that predicates have. Returning to one of our early examples, the Logistics domain from Section 2.1.3, this allows us to define functions such as

```
(free_capacity ?t - truck) - quantity
```

instead of using predicates with the quantity as an extra parameter. Not only does this make the formulation of action effects easier, but it also makes it explicit that the quantity parameter in these predicates is functionally dependent on the other arguments, i.e., that in any state there will be only one value of (`free_capacity ?t`) for each truck `?t`. The addition of general finite-domain state variables does not change the expressivity of the language, in the sense that the complexity of the class of problems it can express is the same.

While recent new Functional STRIPS planners have been developed [Frances and Geffner, 2015, Frances et al., 2017], this extension to PDDL has not been widely adopted. Partly this may be because there are some subtle issues with the typing of object-valued functions that sometimes make their use not as natural as one would hope. Partly it is the chicken-and-egg problem that problem modellers often choose not to use features that are not widely supported by planners, while planner implementors often choose not to support features that are not commonly used in existing PDDL models. As we have seen in several examples in this book, classical planning formulations often use objects to represent a finite range of integers. A small extension that would perhaps make object-valued functions more practically appealing would be a simplified syntax for declaring such finite integer ranges implicitly and allowing the use of arithmetic operators on them, subsequently allowing for more compact encodings.

In the numeric subset of PDDL, which we described in Chapter 4, numeric expressions can only be made up using the four standard arithmetic operators (+, −, ∗, and /). This limitation is somewhat arbitrary: on the one hand, it already leads to a semi-decidable planning problem, while on the other it prevents the formulation of many interesting numeric planning problems which require expressions involving trigonometric operators, square roots, etc. There are already planners that support such operators [e.g., Scala et al., 2016], and adding them to PDDL is, syntactically, a small step. Quantifying numeric operators, such as sum and product, analogous to the logical quantifiers `forall` and `exists`, would also make it easier to write general numeric planning domains, without changing expressivity essentially. Allowing numeric-valued action parameters, known as "control parameters", would be a bigger departure from current numeric and hybrid PDDL. Control parameters allow, for example, an action to be parameterised with an arbitrary amount, or an action with continuous change to be parameterised with a rate. There is current work on extending planners to this kind of problem [e.g., Savas et al., 2016].

MODULAR EXTENSIONS OF PLANNING MODELS

No matter how expressive PDDL becomes, there will always be some aspects of some planning problems that are difficult, or even impossible, to capture in it. In practical planning applications, it is also often the case that planners must interface with specialised or legacy systems that provide "models" of some parts of the problem.

Recognising this, researchers have long looked at the question of how to extend planning languages in a modular way, by integrating "black box" modules that compute, for example, some function, predicate, or even parts of a plan, into planning systems [e.g., Aylett et al., 1998, Currie and Tate, 1991, Frances et al., 2017, Jonsson et al., 1999, to name a few]. A special case that has received particular attention is the integration of robotic motion planning within the action-oriented PDDL style of planning [e.g., Cambon et al., 2009, Lagriffoul et al., 2012, Plaku and Hager, 2010, Srivastava et al., 2014, Toussaint, 2015]. Recent work on general mechanisms for such integration have termed such modules "sematic attachments" [Bernardini et al., 2017b, Dornhege et al., 2009], "planning modulo theories" [Gregory et al., 2012], "factored state models" specifying simulators [Frances et al., 2017] or "secondary models" [Haslum et al., 2018]. There is as yet no agreed form for expressing the inclusion of external modules in PDDL, but several proposals exist.

MODELLING TOOL SUPPORT

The main purpose of modelling planning problems in PDDL is to apply automated planning systems to find solution plans. However, developing and improving other tools that aid in the process of creating PDDL models may also accelerate wider use of PDDL planning.

A syntax checker and plan validator (see `https://github.com/KCL-Planning/VAL` and `https://github.com/patrikhaslum/INVAL`) can be invaluable in the task of testing and debugging a PDDL model. Their usefulness is only somewhat diminished by the fact that validators and planners sometimes implement different interpretations of some aspects of PDDL. Improving the ability of plan validators to explain why a plan fails is a feature that could further ease the modelling, and model debugging, task.

Editors with PDDL-specific features, from syntax highlighting to integration with planners, template-based autofill, and more, can facilitate writing PDDL models in the first place. We have made references to the on-line editor at `http://editor.planning.domains` throughout this book. The PDDL plugin for the VSCode IDE [Long et al., 2018, see also ht `tps://marketplace.visualstudio.com/items?itemName=jan-dolejsi.pddl`] is another recent example. Graphical tools for defining domains, without having to explicitly write them in PDDL, have also been proposed [e.g., Vaquero et al., 2007, see also `https://github.com/tvaquero/itsimple`].

APPENDIX A

Online PDDL Resources

- The PDDL Resources page of IPC 2008: `http://icaps-conference.org/ipc2008/deterministic/PddlResources.html`

  Collects copies of most documents describing versions of PDDL up to 2008.

- PDDL Wikipedia page: `https://en.wikipedia.org/wiki/Planning_Domain_Definition_Language`

- Planning.Domains: `http://planning.domains/`

  Repository of planning domains and problems, on-line PDDL editor and planner, and collection of educational resources.

- PDDL 1.2 [Bacchus, 2000]: `http://ipc00.icaps-conference.org/pddl-subset.ps`

- PDDL 2.1 [Fox and Long, 2003]: `https://www.jair.org/index.php/jair/article/view/10352`

- PDDL 2.2 [Hoffmann and Edelkamp, 2005]: `https://www.jair.org/index.php/jair/article/view/10427`

- PDDL 3.0 [Gerevini et al., 2009]: `https://www.sciencedirect.com/science/article/pii/S0004370208001847`

- PPDDL / FOND Bryce and Buffet [2008]: `http://icaps-conference.org/ipc2008/probabilistic/wiki/images/2/21/Rules.pdf`

- MA-PDDL: `http://www.r3-cop.eu/wp-content/uploads/2013/01/A-Multy-Agent-Extension-of-PDDL3.1.pdf`

- VAL: `https://github.com/KCL-Planning/VAL`

  Includes a PDDL syntax checker, a plan validator, and other tools.

- INVAL: `https://github.com/patrikhaslum/INVAL`

  Includes a plan validator (for non-temporal domains only) and other tools.

Bibliography

Mitchell Ai-Chang, John Bresina, Len Charest, Adam Chase, Jennifer Cheng Jung Hsu, Ari Jonsson, Bob Kanefsky, Paul Morris, Kanna Rajan, Jeffrey Yglesias, Brian G. Chafin, William C. Dias, and Pierre F. Maldague. MAPGEN Planner: Mixed-initiative activity planning for the Mars Exploration Rover mission. *IEEE Intelligent Systems*, 19(1):8–12, 2004. DOI: 10.1109/MIS.2004.1265878 7

Masataro Asai and Alex Fukunaga. Fully automated cyclic planning for large-scale manufacturing domains. In *Proc. of the 24th International Conference on Automated Planning and Scheduling (ICAPS14)*, pages 20–28, 2014. 7

Ruth Aylett, James K. Soutter, Gary J. Petley, and Paul W. H. Chung. AI planning in a chemical plant domain. In *European Conference on AI*, pages 622–626, 1998. 6, 146

Fahiem Bacchus. Subset of PDDL for the AIPS 2000 planning competition. http://ipc00.icaps-conference.org/pddl-subset.ps, 2000. 8, 149

Christer Bäckström. Computational aspects of reordering plans. *Journal of AI Research*, 9:99–137, 1998. DOI: 10.1613/jair.477 40, 42

Javier Barreiro, Matthew Boyce, Minh Do, Jeremy Frank, Michael Iatauro, Tatiana Kichkaylo, Paul Morris, James Ong, Emilio Remolina, Tristan Smith, and David Smith. EUROPA: A platform for AI planning, scheduling, constraint programming, and optimization. In *Proc. of the 4th International Competition on Knowledge Engineering for Planning and Scheduling (ICKEPS)*, 2012. 10

J. Benton, Amanda Jane Coles, and Andrew Coles. Temporal planning with preferences and time-dependent continuous costs. In *Proc. of the 22nd International Conference on Automated Planning and Scheduling, (ICAPS)*, Atibaia, São Paulo, Brazil, June 25–19, 2012. http://www.aaai.org/ocs/index.php/ICAPS/ICAPS12/paper/view/4699. 109

Sara Bernardini, Maria Fox, and Derek Long. Planning the behaviour of low-cost quadcopters for surveillance missions. In *Proc. of the 24th International Conference on Automated Planning and Scheduling (ICAPS)*, pages 445–453, 2017a. 7

Sara Bernardini, Maria Fox, Derek Long, and Chiara Piacentini. Boosting search guidance in problems with semantic attachments. In *Proc. 27th International Conference on Automated Planning and Scheduling (ICAPS)*, pages 29–37, 2017b. 146

Armin Biere, Marijn Heule, Hans van Maaren, and Toby Walsh, Eds. *Handbook of Satisfiability*, vol. 185 of *Frontiers in Artificial Intelligence and Applications*, IOS Press, 2009. 10

Mark S. Boddy, Johnathan Gohde, Thomas Haigh, and Steven A. Sharp. Course of action generation for cyber security using classical planning. In *Proc. of the 15th International Conference on Automated Planning and Scheduling (ICAPS)*, pages 12–21, 2005. 6

Blai Bonet and Hector Geffner. GPT: A tool for planning with uncertainty and partial information. In *Proc. of the Workshop on Planning with Uncertainty and Partial Information (IJCAI'01)*, 2001. 142

George S. Boolos and Richard C. Jeffrey. *Computability and Logic*, 3rd ed., Cambridge University Press, 1980. DOI: 10.1017/cbo9781139164931 99

Roberto Boselli, Mirko Cesarini, Fabio Mercorio, and Mario Mezzanzanica. Planning meets data cleansing. In *Proc. 24th International Conference on Automated Planning and Scheduling (ICAPS)*, 2014. 5

Michael Brenner. A multiagent planning language. In *Proc. ICAPS Workshop on PDDL*, 2003. 143

Michael Brenner. Creating dynamic story plots with continual multiagent planning. In *Proc. of the 24th AAAI Conference on Artificial Intelligence*, pages 1517–1522, 2010. 7

Daniel Bryce and Olivier Buffet. 6th international planning competition: Uncertainty part. *Proc. of the 6th International Planning Competition (IPC'08)*, 2008. 139, 149

Tom Bylander. Complexity results for planning. In *Proc. 12th International Joint Conference on Artificial Intelligence (IJCAI)*, pages 274–279, 1991. 55, 58

Alberto Camacho, Jorge A. Baier, Christian J. Muise, and Sheila A. McIlraith. Synthesizing controllers: On the correspondence between LTL synthesis and non-deterministic planning. In *Advances in Artificial Intelligence—Proc. of the 31st Canadian Conference on Artificial Intelligence (CCAI)*, pages 45–59, 2018. DOI: 10.1007/978-3-319-89656-4_4 11

Stéphane Cambon, Rachid Alami, and Fabien Gravot. A hybrid approach to intricate motion, manipulation and task planning. *International Journal of Robotics Research*, 28(1):104–126, 2009. DOI: 10.1177/0278364908097884 7, 146

Michael Cashmore, Maria Fox, Tom Larkworthy, Derek Long, and Daniele Magazzeni. AUV mission control via temporal planning. In *IEEE International Conference on Robotics and Automation, (ICRA)*, pages 6535–6541, Hong Kong, China, May 31–June 7, 2014. DOI: 10.1109/ICRA.2014.6907823 7

Michael Cashmore, Maria Fox, Derek Long, Daniele Magazzeni, Bram Ridder, Arnau Carrera, Narcís Palomeras, Natália Hurtós, and Marc Carreras. Rosplan: Planning in the robot operating system. In *Proc. of the 25th International Conference on Automated Planning and Scheduling, (ICAPS)*, pages 333–341, 2015. 7

Nathanael Chambers and Dan Jurafsky. Unsupervised learning of narrative schemas and their participants. In *Proc. of the 47th Annual Meeting of the Association for Computational Linguistics and the 4th International Joint Conference on Natural Language Processing (ACL-IJCNLP)*, pages 602–610, 2009. DOI: 10.3115/1690219.1690231 8

Hsueh-Min Chang and Von-Wun Soo. Simulation-based story generation with a theory of mind. In *Proc. Artificial Intelligence and Interactive Digital Entertainment Conference (AIIDE)*, pages 16–21, 2008. 7

Steve Chien, Gregg Rabideau, Russel Knight, Rob Sherwood, Barbara Engelhardt, Darren Mutz, Tara Estlin, Benjamin Smith, Forest Fisher, Anthony Barrett, G. Stebbins, and Daniel Tran. ASPEN—automating space mission operations using automated planning and scheduling. In *Proc. 6th International Symposium on Technical Interchange for Space Mission Operations*, 2000. 10

Edmund M. Clarke, Orna Grumberg, and Doron Peled. *Model Checking*, MIT Press, 1993. DOI: 10.1007/978-3-642-61455-2_16 11, 48

Matthew Crosby, Ronald P. A. Petrick, Francesco Rovida, and Volker Krüger. Integrating mission and task planning in an industrial robotics framework. In *Proc. of the 27th International Conference on Automated Planning and Scheduling (ICAPS)*, pages 471–479, 2017. 7

Ken Currie and Austin Tate. O-Plan: The open planning architecture. *Artificial Intelligence*, 52:49–86, 1991. DOI: 10.1016/0004-3702(91)90024-e 146

George B. Dantzig. On the significance of solving linear programming problems with some integer variables. *Econometrica, Journal of the Econometric Society*, pages 30–44, 1960. DOI: 10.2307/1905292 10

George B. Dantzig and John H. Ramser. The truck dispatching problem. *Management Science*, 6(1):80–91, 1959. DOI: 10.1287/mnsc.6.1.80 5

Giuseppe De Giacomo, Fabrizio Maria Maggi, Andrea Marrella, and Sebastian Sardina. Computing trace alignment against declarative process models through planning. In *Proc. of the 26th International Conference on Automated Planning and Scheduling (ICAPS'16)*, pages 367–375, 2016. 5

Christian Dornhege, Patrick Eyerich, Thomas Keller, Sebastian Trüg, Michael Brenner, and Bernhard Nebel. Semantic attachments for domain-independent planning systems. In *Proc.*

19th International Conference on Automated Planning and Scheduling (ICAPS), 2009. DOI: 10.1007/978-3-642-25116-0_9 146

Michael Drexl. Rich vehicle routing in theory and practice. *Logistics Research*, 5:47–63, 2012. DOI: 10.1007/s12159-012-0080-2 5

Stefan Edelkamp and Jörg Hoffmann. PDDL2.2: The language for the classical part of IPC-4. In *4th International Planning Competition Booklet*, pages 2–6, 2004. http://ipc.icaps-conference.org/ 9, 118

Stefan Edelkamp, Peter Kissmann, and Álvaro Torralba. BDDs strike back (in AI planning). In *Proc. of the 29th AAAI Conference on Artificial Intelligence*, pages 4320–4321, 2015. http://www.aaai.org/ocs/index.php/AAAI/AAAI15/paper/view/9834 60

Robert T. Effinger, Brian C. Williams, Gerard Kelly, and Michael Sheehy. Dynamic controllability of temporally-flexible reactive programs. In *Proc. 19th International Conference on Automated Planning and Scheduling (ICAPS)*, 2009. http://aaai.org/ocs/index.php/ICAPS/ICAPS09/paper/view/739 10

Esra Erdem and Elisabeth Tillier. Genome rearrangement and planning. In *Proc. 20th National Conference on AI (AAAI'05)*, 2005. 6

Kutluhan Erol, Dana S. Nau, and V. S. Subrahmanian. Complexity, decidability and undecidability results for domain-independent planning: A detailed analysis. *Technical Report CS-TR-2797*, Computer Science Department, University of Maryland, 1991. DOI: 10.1016/0004-3702(94)00080-k 58

Richard E. Fikes and Nils J. Nilsson. STRIPS: A new approach to the application of theorem proving to problem solving. *Artificial Intelligence*, 2:189–208, 1971. DOI: 10.1016/0004-3702(71)90010-5 6, 8

José E. Flórez, Álvaro Torralba Arias de Reyna, Javier García, Carlos Linares López, Angel García Olaya, and Daniel Borrajo. Planning multi-modal transportation problems. In *Proc. 21st International Conference on Automated Planning and Scheduling (ICAPS)*, pages 66–73, 2011. 5

Maria Fox and Derek Long. PDDL2.1: An extension to PDDL for expressing temporal planning domains. *Journal of AI Research*, 20:61–124, 2003. DOI: 10.1613/jair.1129 8, 91, 92, 93, 103, 114, 116, 119, 122, 149

Maria Fox and Derek Long. Modelling mixed discrete-continuous domains for planning. *Journal of AI Research*, 27:235–297, 2006. DOI: 10.1613/jair.2044 8, 123, 137

Maria Fox, Richard Howey, and Derek Long. Val's progress: The automatic validation tool for PDDL2.1 used in the international planning competition. In *IPC at ICAPS*, 2003. 127

Maria Fox, Derek Long, and Keith Halsey. An investigation into the expressive power of PDDL2.1. In *Proc. of the 15th European Conference on Artificial Intelligence, (ECAI)*, pages 338–342, 2004. 113

Maria Fox, Richard Howey, and Derek Long. Validating plans in the context of processes and exogenous events. In *Proc. of the 20th National Conference on Artificial Intelligence and the 7th Innovative Applications of Artificial Intelligence Conference*, pages 1151–1156, Pittsburgh, PA, July 9–13, 2005. 123, 125, 134

Maria Fox, Derek Long, and Daniele Magazzeni. Plan-based policies for efficient multiple battery load management. *Journal of Artificial Intelligence Research (JAIR)*, 44:335–382, 2012. DOI: 10.1613/jair.3643 6, 129

Guillem Frances and Hector Geffner. Modeling and computation in planning: Better heuristics from more expressive languages. In *Proc. of the 25th International Conference on Planning and Scheduling (ICAPS)*, 2015. 145

Guillem Frances, Miquel Ramírez Jávega, Nir Lipovetzky, and Hector Geffner. Purely declarative action descriptions are overrated: Classical planning with simulators. In *Proc. of 26th International Joint Conference on Artificial Intelligence (IJCAI)*, pages 4294–4301, 2017. DOI: 10.24963/ijcai.2017/600 145, 146

Jeremy Frank and Ari K. Jonsson. Constraint-based attribute and interval planning. *Constraints*, 8:339–364, 2003. DOI: 10.1023/A:1025842019552 10

Martin Gardner. Mathematical games. *Scientific American Magazine*, January 1973. DOI: 10.1038/scientificamerican0169-116 94

M. R. Garey and D. S. Johnson. *Computers and Intractability: A Guide to the Theory of NP-Completeness*, Freeman, 1979. DOI: 10.1137/1024022 53

Héctor Geffner. Functional strips: A more flexible language for planning and problem solving. In *Logic-Based Artificial Intelligence*, pages 187–209, Springer, 2000. DOI: 10.1007/978-1-4615-1567-8_9 145

Hector Geffner and Blai Bonet. *A Concise Introduction to Models and Methods for Automated Planning*, Morgan & Claypool, 2013. DOI: 10.2200/s00513ed1v01y201306aim022 xvii

Alfonso Gerevini, Patrik Haslum, Derek Long, Alessandro Saetti, and Yannis Dimopoulos. Deterministic planning in the 5th international planning competition: PDDL3 and experimental evaluation of the planners. *Artificial Intelligence*, 173(5–6):619–668, 2009. DOI: 10.1016/j.artint.2008.10.012 9, 77, 82, 149

Pablo Gervás. Computational approaches to storytelling and creativity. *AI Magazine*, 30(3):49–62, 2009. DOI: 10.1609/aimag.v30i3.2250 7

Malik Ghallab and Hervé Laruelle. Representation and control in IxTeT, a temporal planner. In *Proc. 2nd International Conference on AI Planning Systems (AIPS'94)*, pages 61–67, 1994. 10, 103

Malik Ghallab, Dana Nau, and Paolo Traverso. *Automated Planning: Theory and Practice*, Morgan Kaufmann Publishers, 2004. xvii

Kamalesh Ghosh, Pallab Dasgupta, and S. Ramesh. Automated planning as an early verification tool for distributed control. *Journal of Automated Reasoning*, 54:31–68, 2015. DOI: 10.1007/s10817-014-9313-1 11

Fausto Giunchiglia and Paolo Traverso. Planning as model checking. In *Recent Advances in AI Planning, 5th European Conference on Planning, (ECP'99) Proceedings*, pages 1–20, Durham, UK, September 8–10, 1999. DOI: 10.1007/10720246_1 11

Robert P. Goldman and Ugur Kuter. Measuring plan diversity: Pathologies in existing approaches and A new plan distance metric. In *Proc. of the 29th AAAI Conference on Artificial Intelligence*, pages 3275–3282, 2015. 8

Cordell Green. Applications of theorem proving to problem solving. In *Proc. International Joint Conference on Artificial Intelligence (IJCAI)*, pages 219–240, 1969. DOI: 10.21236/ada459656 8

Peter Gregory, Derek Long, Maria Fox, and Christopher Beck. Planning modulo theories: Extending the planning paradigm. In *Proc. 22nd International Conference on Automated Planning and Scheduling (ICAPS)*, pages 65–73, 2012. 146

S. Hannenhalli and P. A. Pevzner. Transforming cabbage into turnip (polynomial algorithm for sorting signed permutations by reversals). In *Proc. 27th ACM-SIAM Symposium on the Theory of Computing (STOC'95)*, pages 178–189, 1995. DOI: 10.1145/300515.300516 44

Patrik Haslum. Computing genome edit distances using domain-independent planning. In *Scheduling and Planning Applications Workshop (ICAPS'11)*, 2011. 6, 45

Patrik Haslum. Narrative planning: Compilations to classical planning. *Journal of AI Research*, 44:383–395, 2012. http://www.jair.org/papers/paper3602.html DOI: 10.1613/jair.3602 7

Patrik Haslum and Alban Grastien. Diagnosis as planning: Two case studies. In *Scheduling and Planning Applications Workshop (ICAPS'11)*, 2011. 5

Patrik Haslum, Franc Ivankovic, Miquel Ramírez, Dan Gordon, Sylvie Thiébaux, Vikas Shivashankar, and Dana S. Nau. Extending classical planning with state constraints: Heuristics and search for optimal planning. *Journal of AI Research*, 62:373–431, 2018. DOI: 10.1613/jair.1.11213 146

Malte Helmert. Decidability and undecidability results for planning with numerical state variables. In *Proc. 6th International Conference on Artificial Intelligence Planning and Scheduling (AIPS)*, pages 303–312, 2002. 98, 100

Malte Helmert and Hauke Lasinger. The Scanalyzer domain: Greenhouse logistics as a planning problem. In *Proc. of the 20th International Conference on Automated Planning and Scheduling (ICAPS'10)*, pages 234–237, 2010. 5

Malte Helmert and Gabriele Röger. How good is almost perfect? In *Proc. 23nd AAAI Conference on Artificial Intelligence (AAAI'08)*, pages 944–949, 2008. 60

Malte Helmert, Mihn Do, and Ioannis Refanidis. Changes in PDDL 3.1. http://icaps-conference.org/ipc2008/deterministic/PddlExtension.html, 2008. 9, 84, 145

C. A. R. Hoare. *Communicating Sequential Processes*, Prentice Hall, 1985. DOI: 10.1007/978-3-662-09507-2_19 48

Jörg Hoffmann. Simulated penetration testing: From "Dijkstra" to "Turing Test++". In *Proc. of the 25th International Conference on Automated Planning and Scheduling (ICAPS)*, pages 364–372, 2015. 6

Jörg Hoffmann and Ronen I. Brafman. Conformant planning via heuristic forward search: A new approach. *Artificial Intelligence*, 170(6–7):507–541, 2006. DOI: 10.1016/j.artint.2006.01.003 142

Jörg Hoffmann and Stefan Edelkamp. The deterministic part of IPC-4: An overview. *Journal of AI Research*, 24:519–579, 2005. DOI: 10.1613/jair.1677 149

Richard Howey, Derek Long, and Maria Fox. VAL: Automatic plan validation, continuous effects and mixed initiative planning using PDDL. In *16th IEEE International Conference on Tools with Artificial Intelligence (ICTAI)*, pages 294–301, Boca Raton, FL, November 15–17, 2004. DOI: 10.1109/ictai.2004.120 126, 134

M. Jongerden and B. Haverkort. Which battery model to use? *IET Software*, 3(6):445–457, 2009. DOI: 10.1049/iet-sen.2009.0001 129, 131

Ari K. Jonsson, Paul H. Morris, and Nicola Muscettola Kanna Rajan. Next generation remote agent planner. In *Proc. of the 5th International Symposium on Artificial Intelligence, Robotics and Automation in Space (iSAIRAS)*, 1999. 146

Ari K. Jonsson, Paul Morris, Nicola Muscettola, Kanna Rajan, and Ben Smith. Planning in interplanetary space: Theory and practice. In *Proc. 5th International Conference on Artificial Intelligence Planning and Scheduling (AIPS'00)*, pages 177–186, 2000. 7

Henry A. Kautz and Bart Selman. Pushing the envelope: Planning, propositional logic and stochastic search. In *Proc. of the 13th National Conference on Artificial Intelligence and 8th Innovative Applications of Artificial Intelligence Conference, (AAAI)*, vol. 2, pages 1194–1201, Portland, OR, August 4–8, 1996. 11

Emil Keyder and Hector Geffner. Soft goals can be compiled away. *Journal of Artificial Intelligence Research*, 36:547–556, 2009. DOI: 10.1613/jair.2857 82

Philip Kilby, Ignasi Abio, Daniel Guimarans, Daniel Harabor, Patrik Haslum, Valentin Mayer-Eichberger, Fazlul Siddiqui, Sylvie Thiebaux, and Tommaso Urli. There's more than one way to solve a long-haul transportation problem. In *Vehicle Routing and Logistics (VeRoLog)*, page 22, 2015. 5, 24

Jana Koehler and Daniel Ottiger. An AI-based approach to destination control in elevators. *AI Magazine*, 23(2):59–78, 2002. 5

Daniel L. Kovacs. A multi-agent extension to PDDL3.1. In *Proc. of the Workshop on the International Planning Competition (ICAPS)*, 2012. http://www.r3-cop.eu/wp-content/uploads/2013/01/A-Multy-Agent-Extension-of-PDDL3.1.pdf 10, 143, 144

Fabien Lagriffoul, Dimitar Dimitrov, Alessandro Saffiotti, and Lars Karlsson. Constraint propagation on interval bounds for dealing with geometric backtracking. In *IEEE/RSJ International Conference on Intelligent Robots and Systems (IROS)*, pages 957–964, 2012. DOI: 10.1109/iros.2012.6385972 7, 146

Steven M. LaValle. *Planning Algorithms*, Cambridge University Press, 2006. DOI: 10.1017/cbo9780511546877 7

Hector J. Levesque, Raymond Reiter, Yves Lesperance, Fangzhen Lin, and Richard B. Scherl. GOLOG: A logic programming language for dynamic domains. *Journal of Logic Programming*, 31:59–83, 1997. DOI: 10.1016/s0743-1066(96)00121-5 10

BoonPing Lim, Menkes van den Briel, Sylvie Thiébaux, Scott Backhaus, and Russell Bent. HVAC-aware occupancy scheduling. In *Proc. of the 29th AAAI Conference on Artificial Intelligence*, pages 679–686, Austin, TX, January 25–30, 2015. http://www.aaai.org/ocs/index.php/AAAI/AAAI15/paper/view/9406 6

Derek Long, Jan Dolejsi, and Maria Fox. Building support for PDDL as a modelling tool. In *Proc. of the ICAPS Workshop on Knowledge Engineering for Planning and Scheduling*, pages 78–82, 2018. http://icaps18.icaps-conference.org/fileadmin/alg/conferences/icaps18/workshops/workshop04/docs/KEPS-proceedings.pdf 146

Jorge Lucangeli Obes, Carlos Sarraute, and Gerardo Richarte. Attack planning in the real world. In *2nd Workshop on Security and Artificial Intelligence*, 2010. 6

J. Manwell and J. McGowan. Lead acid battery storage model for hybrid energy systems. *Solar Energy*, 50(1):399–405, 1993. DOI: 10.1016/0038-092x(93)90060-2 129

Bart Massey. Directions in planning: Understanding the flow of time in planning. Ph.D. thesis, University of Oregon, 1999. 58

Mausam and Andrey Kolobov. *Planning with Markov Decision Processes: An AI Perspective*, Morgan & Claypool, 2012. DOI: 10.2200/S00426ED1V01Y201206AIM017 9

Ernst W. Mayr. An algorithm for the general Petri net reachability problem. In *Proc. 13th Annual ACM Symposium on the Theory of Computing (STOC)*, pages 238–246, 1981. DOI: 10.1145/800076.802477 100

Drew McDermott. The 1998 AI planning systems competition. *AI Magazine*, 21(2):35–55, 2000. http://www.aaai.org/ojs/index.php/aimagazine/article/view/1506 8

Drew McDermott, Malik Ghallab, Adele Howe, Craig Knoblock, Ashwin Ram, Manuela Veloso, Daniel Weld, and David Wilkins. PDDL—the planning domain definition language. *Technical Report CVC TR-98-003/DCS TR-1165*, Yale Center for Computational Vision and Control, 1998. 8, 15, 42, 43

Conor McGann, Frederic Py, Kanna Rajan, Hans Thomas, Richard Henthorn, and Robert S. McEwen. A deliberative architecture for AUV control. In *IEEE International Conference on Robotics and Automation (ICRA'08)*, pages 1049–1054, 2008. DOI: 10.1109/robot.2008.4543343 7

Shuwa Miura and Alex Fukunaga. Automatic extraction of axioms for planning. In *Proc. of the 27th International Conference on Automated Planning and Scheduling (ICAPS)*, pages 218–227, Pittsburgh, PA, June 18–23, 2017. https://aaai.org/ocs/index.php/ICAPS/ICAPS17/paper/view/15730 74

Christian Muise and Nir Lipovetzky. Unsolvability international planning competition. https://unsolve-ipc.eng.unimelb.edu.au/, 2016. 11

Christian Muise, Christopher Beck, and Sheila McIlraith. Optimal partial-order plan relaxation via MaxSAT. *Journal of AI Research*, 57:113–149, 2016. DOI: 10.1613/jair.5128 40

Nicola Muscettola. Integrating planning and scheduling. In M. Zweben and M. Fox, Eds., *Intelligent Scheduling*, Morgan Kaufmann, 1994. 10, 103

Dana S. Nau, Tsz-Chiu Au, Okhtay Ilghami, Ugur Kuter, Héctor Muñoz-Avila, J. William Murdock, Dan Wu, and Fusun Yaman. Applications of SHOP and SHOP2. *IEEE Intelligent Systems*, 20(2):34–41, 2005. DOI: 10.1109/MIS.2005.20 10

Bernhard Nebel and Christer Bäckström. On the computational complexity of temporal projection, planning, and plan validation. *Artificial Intelligence*, 66(1):125–160, 1994. 42

Bernhard Nebel. On the compilability and expressive power of propositional planning formalisms. *Journal of AI Research*, 12:271–315, 2000. DOI: 10.1613/jair.735 79, 80

Nicholas Nethercote, Peter J. Stuckey, Ralph Becket, Sebastian Brand, Gregory J. Duck, and Guido Tack. Minizinc: Towards a standard CP modelling language. In *International Conference on Principles and Practice of Constraint Programming*, pages 529–543, Springer, 2007. DOI: 10.1007/978-3-540-74970-7_38 10

Fabio Patrizi, Nir Lipovetzky, Giuseppe De Giacomo, and Hector Geffner. Computing infinite plans for LTL goals using a classical planner. In *IJCAI*, pages 2003–2008, 2011. DOI: 10.5591/978-1-57735-516-8/IJCAI11-334 77

Fabio Patrizi, Nir Lipovetzky, and Hector Geffner. Fair LTL synthesis for non-deterministic systems using strong cyclic planners. In *Proc. of the 23rd International Joint Conference on Artificial Intelligence (IJCAI)*, pages 2343–2349, Beijing, China, August 3–9, 2013. 11

Edwin Pednault. ADL: Exploring the middle ground between STRIPS and the situation calculus. In *Proc. 1st International Conference on Principles of Knowledge Representation and Reasoning (KR'89)*, pages 324–332, 1989. 63

J. Scott Penberthy and Daniel S. Weld. Temporal planning with continuous change. In *Proc. 12th National Conference on Artificial Intelligence (AAAI'94)*, pages 1010–1015, 1994. 103

Chiara Piacentini, Varvara Alimisis, Maria Fox, and Derek Long. An extension of metric temporal planning with application to AC voltage control. *Artificial Intelligence*, 229:210–245, 2015. DOI: 10.1016/j.artint.2015.08.010 6

Chiara Piacentini, Daniele Magazzeni, Derek Long, Maria Fox, and Chris Dent. Solving realistic unit commitment problems using temporal planning: Challenges and solutions. In Amanda Jane Coles, Andrew Coles, Stefan Edelkamp, Daniele Magazzeni, and Scott Sanner, Eds., *Proc. of the 26th International Conference on Automated Planning and Scheduling (ICAPS)*, pages 421–430, AAAI Press, London, UK, June 12–17, 2016. http://www.aaai.org/ocs/index.php/ICAPS/ICAPS16/paper/view/12992 6

Erion Plaku and Gregory D. Hager. Sampling-based motion and symbolic action planning with geometric and differential constraints. In *Proc. IEEE International Conference on Robotics and Automation (ICRA)*, pages 5002–5008, 2010. DOI: 10.1109/robot.2010.5509563 7, 146

Amir Pnueli. The temporal logic of programs. In *18th Annual Symposium on Foundations of Computer Science*, pages 46–57, Providence, RI, October 31–November 1, 1977. DOI: 10.1109/SFCS.1977.32 77

Julie Porteous, Fred Charles, and Marc Cavazza. NetworkING: Using character relationships for interactive narrative generation. In *International Conference on Autonomous Agents and Multi-Agent Systems (AAMAS)*, pages 595–602, 2013. 7

Julie Porteous, Alan Lindsay, Jonathon Read, Mark Truran, and Marc Cavazza. Automated extension of narrative planning domains with antonymic operators. In *Proc. of the 2015 International Conference on Autonomous Agents and Multiagent Systems (AAMAS)*, pages 1547–1555, 2015. 8

Martin L. Puterman. *Markov Decision Processes: Discrete Stochastic Dynamic Programming*. Wiley, 1994. DOI: 10.2307/1269932 9

Jussi Rintanen. Heuristics for planning with SAT. In *Proc. 16th International Conference on Principles and Practice of Constraint Programming (CP'10)*, vol. 6308 of *LNCS*, pages 414–428, 2010. DOI: 10.1007/978-3-642-15396-9_34 60

Francesca Rossi, Peter van Beek, and Toby Walsh, Eds. *Handbook of Constraint Programming*, vol. 2 of *Foundations of Artificial Intelligence*. Elsevier, 2006. 10

Stuart J. Russel and Peter Norvig. *Artificial Intelligence: A Modern Approach*, 3rd ed., Prentice Hall, 2011. xvii

Scott Sanner. Relational dynamic influence diagram language (RDDL): Language description. http://users.cecs.anu.edu.au/~ssanner/IPPC_2011/RDDL.pdf, 2011. 9, 141

Emre Savas, Maria Fox, Derek Long, and Daniele Magazzeni. Planning using actions with control parameters. In *Proc. 22nd European Conference on Artificial Intelligence (ECAI)*, pages 1185–1193, 2016. DOI: 10.3233/978-1-61499-672-9-1185 86, 146

Walter J. Savitch. Relationships between nondeterministic and deterministic tape complexities. *Journal of Computer and Systems Sciences*, 4(2):177–192, 1970. DOI: 10.1016/s0022-0000(70)80006-x 55

Enrico Scala, Patrik Haslum, Sylvie Thiébaux, and Miquel Ramírez. Interval-based relaxation for general numeric planning. In *22nd European Conference on Artificial Intelligence (ECAI)*, pages 655–663, 2016. 145

Thomas J. Schaefer. The complexity of satisfiability problems. In *Proc. of the 10th Annual ACM Symposium on Theory of Computing*, pages 216–226, San Diego, CA, May 1–3, 1978. DOI: 10.1145/800133.804350 10

Avirup Sil and Alexander Yates. Extracting STRIPS representations of actions and events. In *Recent Advances in Natural Language Processing (RANLP)*, pages 1–8, 2011. 8

David E. Smith and Daniel S. Weld. Temporal planning with mutual exclusion reasoning. In *Proc. 16th International Joint Conference on Artificial Intelligence (IJCAI'99)*, pages 326–333, 1999. 103

David E. Smith, Jeremy Frank, and William Cushing. The ANML language. In *Proc. of the Workshop on Knowledge Engineering for Planning and Scheduling (ICAPS)*, 2008. `http://kt iml.mff.cuni.cz/~bartak/KEPS2008/download/paper07.pdf` 10

Biplav Srivastava, Subbarao Kambhampati, Tuan Nguyen, Minh Binh Do, Alfonso Gerevini, and Ivan Serina. Domain independent approaches for finding diverse plans. In *Proc. 20th International Conference on Artificial Intelligence (IJCAI)*, pages 2016–2022, 2007. 8

Siddharth Srivastava, Eugene Fang, Lorenzo Riano, Rohan Chitnis, Stuart Russel, and Pieter Abbeel. Combined task and motion planning through an extensible planner-independent interface layer. In *Proc. IEEE International Conference on Robotics and Automation (ICRA)*, pages 639–646, 2014. DOI: 10.1109/icra.2014.6906922 7, 146

Martin Suda. Duality in STRIPS planning. In *Proc. of the Workshop on Heuristics and Search for Domain-Independent Planning (ICAPS)*, pages 21–27, 2016. 58

Austin Tate, Brian Drabble, and Richard Kirby. O-Plan2: An open architecture for command, planning and control. In M. Zweben and M. Fox, Eds., *Intelligent Scheduling*, pages 213–239, Morgan Kaufmann, 1994. 10

Sylvie Thiébaux, Jörg Hoffmann, and Bernhard Nebel. In defense of PDDL axioms. *Artificial Intelligence*, 168(1–2):38–69, 2005. DOI: 10.1016/j.artint.2005.05.004 72, 74, 82

Marc Toussaint. Logic-geometric programming: An optimization-based approach to combined task and motion planning. In *Proc. 24th International Joint Conference on AI (IJCAI)*, pages 1930–1936, 2015. 7, 146

Mauro Vallati, Daniele Magazzeni, Bart De Schutter, Lukás Chrpa, and Thomas Leo Mc-Cluskey. Efficient macroscopic urban traffic models for reducing congestion: A PDDL+ planning approach. In Dale Schuurmans and Michael P. Wellman, Eds., *Proc. of the 30th Conference on Artificial Intelligence*, pages 3188–3194, AAAI Press, Phoenix, AZ, February 12–17, 2016. `http://www.aaai.org/ocs/index.php/AAAI/AAAI16/paper/view/11985` 6

Peter van Beek and Xinguang Chen. CPLan: A constraint programming approach to planning. In *Proc. of the 16th National Conference on Artificial Intelligence and 11th Conference on Innovative Applications of Artificial Intelligence*, pages 585–590, Orlando, FL, July 18–22, 1999. 11

Tiago Stegun Vaquero, Victor Romero, Flavio Tonidandel, and José Reinaldo Silva. itSIMPLE 2.0: An integrated tool for designing planning domains. In *Proc. of the 17th International Conference on Automated Planning and Scheduling (ICAPS)*, pages 336–343, 2007. 147

Steven A. Vere. Planning in time: Windows and durations for activities and goals. *IEEE Transactions on Pattern Analysis and Machine Intelligence*, 5:246–267, 1983. DOI: 10.1109/tpami.1983.4767389 103

Stephen G. Ware and Robert Michael Young. Glaive: A state-space narrative planner supporting intentionality and conflict. In *Proc. of the 10th AAAI Conference on Artificial Intelligence and Interactive Digital Entertainment (AIIDE)*, 2014. 7

David E. Wilkins. Hierarchical planning: Definition and implementation. In *Proc. 7th European Conference on Artificial Intelligence (ECAI)*, pages 659–671, 1986. 10

Brian C. Williams and Michel D. Ingham. Model-based programming: Controlling embedded systems by reasoning about hidden state. In *Proc. of Principles and Practice of Constraint Programming (CP)*, pages 508–524, 2002. DOI: 10.1007/3-540-46135-3_34 10

Joanna H. Wimpenny, Alex A. S. Weir, Lisa Clayton, Christian Rutz, and Alex Kacelnik. Cognitive processes associated with sequential tool use in new caledonian crows. *PLOS ONE*, 2009. http://dx.plos.org/10.1371/journal.pone.0006471 DOI: 10.1371/journal.pone.0006471 1

Laurence A. Wolsey. *Integer Programming*. Wiley, 1998. DOI: 10.1002/9781118627372.ch2 10

Håkan L. S. Younes and Michael L. Littman. PPDDL 1.0: An extension to PDDL for expressing planning domains with probabilistic effects. *Technical Report CMU-CS-04-167*, Carnegie Mellon University, 2004. 9, 140

Håkan L. S. Younes and Reid G. Simmons. On the role of ground actions in refinement planning. In *Proc. 6th International Conference on Artificial Intelligence Planning and Scheduling (AIPS'02)*, pages 54–61, 2002. 59

Authors' Biographies

PATRIK HASLUM

Patrik Haslum received his Ph.D. in computer science from Linköping University in 2006, and he is currently at the Australian National University in Canberra. His main area of research is AI planning, with a focus on problem modelling and bridging planning and optimisation.

NIR LIPOVETZKY

Nir Lipovetzky is a Lecturer at the School of Computing and Information Systems at The University of Melbourne. He received his Ph.D. in Computer Science from Universitat Pompeu Fabra, Barcelona. His main research area is Automated Planning, Search, Optimization, and Operations Research. He received the International Conference on Automated Planning and Scheduling (ICAPS) best dissertation award in 2013 for his work under the supervision of Prof. Hector Geffner, the ICAPS best paper award in 2015, and the winner and runner-up awards in two tracks of the International Planning Competition (IPC) in 2018. He served as program chair for ICAPS 2019.

DANIELE MAGAZZENI

Daniele Magazzeni is an Associate Professor at King's College London. He received his Ph.D. in Computer Science from University of L'Aquila in 2009. His research interests are in safe, trusted, and explainable AI, with a particular focus on AI planning for robotics and autonomous systems, and human-AI teaming.

He is the President-Elect of the Executive Council of the International Conference on Automated Planning and Scheduling (ICAPS).

CHRISTIAN MUISE

Christian Muise is a Research Scientist at the MIT-IBM Watson AI Lab, where he researches data-driven techniques for inducing behavioural insight and leads a project devising next generation dialogue agents. Prior to this, he was a Research Fellow with the MERS group at MIT's Computer Science and Artificial Intelligence Laboratory studying decision making under uncertainty, and prior to his time at MIT, Christian was a postdoctoral fellow at the University of Melbourne's Agentlab studying techniques for multi-agent planning and human-agent collaboration. Christian completed his Ph.D. at the University of Toronto with the Knowledge Representation and Reasoning Group in the area of Automated Planning. He is the core developer and active maintainer of the Planning.Domains initiative and has a history of promoting modeling techniques for automated planning.

Index

Printed in the United States
by Baker & Taylor Publisher Services